Physics and Chemistry in Space   Vol. 18
Planetology

Edited by L. J. Lanzerotti, Murray Hill and
D. Stöffler, Münster

M. Gadsden · W. Schröder

Noctilucent Clouds

With 65 Figures

Springer-Verlag
Berlin Heidelberg New York
London Paris Tokyo Hong Kong

Prof. Dr. MICHAEL GADSDEN
Department of Natural Philosophy
University of Aberdeen
Aberdeen AB9 2UE
Scotland

Dr. WILFRIED SCHRÖDER
Hechelstraße 8
2820 Bremen-Rönnebeck
FRG

ISBN 3-540-50685-3 Springer-Verlag Berlin Heidelberg New York
ISBN 0-387-50685-3 Springer-Verlag New York Berlin Heidelberg

Library of Congress Cataloging-in-Publication Data. Gadsden, Michael, 1933- Noctilucent clouds / Michael Gadsden, Wilfried Schröder. p. cm.–(Physics and chemistry in space ; v. 18) Includes index. 1. Noctilucent clouds. I. Schröder, Wilfried. II. Title. III. Series. QC801.P46 vol. 18 [QC976.N6] 530'.0919 s–dc20 [551.57'6]   896277

© Springer-Verlag Berlin Heidelberg 1989
Printed in the United States of America

Typesetting: International Typesetters Inc., Makati, Philippines
2132/3145-543210 – Printed on acid-free paper

Preface

An atmospheric phenomenon is considered as 'explained'
when we have succeeded in deducing it on the basis of
accepted principles of physics.
(H. Ertel, Methods and problems of dynamical meteorology, p. 1)

Until recently, noctilucent clouds were regarded merely as a curious atmospheric phenomenon, the occurrence of which aroused only limited scientific attention. However, in the last two decades the interest they have been given has markedly increased. The clouds, usually pale blue in colour, may be seen on a clear night at high latitudes. Typical examples are illustrated in the book. Clouds looking like these in daytime would be classified as cirrostratus. What sets noctilucent clouds apart is their occurrence in the middle of the night, their very obvious pale blue colour, and their disappearance into the dawn close to the onset of civil twilight when the Sun is 6° below the horizon.

Noctilucent clouds were first recognized as being set apart from ordinary clouds in 1884/1885 and in a series of sightings that followed their return in the summer of 1885. That year marked the beginning of observations and the interpretation of twilight phenomena. The impetus came from the extraordinary Krakatoa eruption, which was not only one of the most spectacular volcanic eruptions in recorded history, but which also turned out to be a startling event in atmospheric science with repercussions even in our time. The uncommon twilight phenomena set a worldwide observational program into motion. Some of the earlier scientific ideas (dating from 1887/1889) from international collaborative work on noctilucent clouds were not fully realized until the International Geophysical Year (1957–1958) and the International Years of Quiet Sun (1964–1965). Not until 1970 were observations standardized by the World Meteorological Organization.

Our book reviews and comments upon research into noctilucent clouds, the historical development, and the current directions that studies are taking. While the volume is self-contained, some sections offer much more difficult reading than others. The reader should feel free to pass over those portions that least interest him, accepting on faith the validity of such intermediate results that find subsequent application. For the more critical reader, an attempt has been made to indicate the basic justification for the various analytical procedures, although little attention has been given to mathematical rigor in the formal sense.

The comprehensive bibliography over of 400 references covers publications from 1648–1989. We believe that the reference list is suitable for supplementing further research and an advanced course in upper atmospheric physics. The chapters have been prepared by both authors: the final version of Chapter 1, and Chapters 4–6, 8, 9 were prepared by MG; Chapters 2, 7, 11 by WS, and Chapters 3 and 10 jointly by both authors. The aim of our work is to provide a useful source book (including different aspects of upper atmospheric physics) for further studies of noctilucent clouds and mesospheric research while at the same time instructing the casual reader and observer.

We are pleased to thank our many colleagues whose suggestions have helped to optimize the presentation of current ideas in this monograph. Dr. L.J. Lanzerotti and Dr. W. Engel have been very helpful in turning the manuscript into a real book.

Aberdeen University   MICHAEL GADSDEN
Aberdeen, Scotland, May 1989   WILFRIED SCHRÖDER

Contents

1 Noctilucent Clouds 1

1.1 Introduction 1
1.2 How, When and Where Noctilucent Clouds Are Seen 1
1.3 Amateur Observations 4
1.4 Cloud Types 5
1.5 Structure of the Upper Atmosphere 10

2 History 13

2.1 Introduction 13
2.2 The Discovery of the "Shining Night-Clouds" 14
2.3 Measurements of Noctilucent Clouds 18
2.4 The Middle Period of Noctilucent Cloud Research 19

3 Observations from Ground Level 23

3.1 Introduction 23
3.2 The Geometry of Twilight Scattering 23
3.3 Latitude of Observation 27
3.4 Absorption of Light in the Atmosphere 28
3.5 Height of Noctilucent Clouds 30
3.6 Drift Motions 36
3.7 Wave Structure 38

4 Spectrophotometry 43

4.1 Introduction 43
4.2 Spectroscopic Observations 44
4.3 Spectrophotometry from Ground Level 45
4.4 Rocket-Borne Photometers 46
4.5 Spectrophotometry from Satellites 49
4.6 Conclusions About Cloud Particle Sizes 56

5 Polarimetry 58

5.1 Introduction 58
5.2 Polarization by Scattering 60
5.3 Measurement of Polarized Light 62

5.4   Polarization Measured from Ground Level   65
5.5   Measurements of Polarization from Rockets   72
5.6   Conclusions About Cloud Particle Sizes   74

6   Rocket-Borne Sampling   75

6.1   Introduction   75
6.2   Flights over Sweden in 1962 and 1967   76
6.3   Flights over Sweden in 1970 and 1971   81
6.4   Flights over Canada in 1968 and 1970   82
6.5   Collectors Flown by Max-Planck-Institut Researchers,
         1968 to 1971   85
6.6   Conclusions About Cloud Particle Sizes   86

7   Variation of Occurrence   87

7.1   Introduction   87
7.2   Sunspot Cycle   87
7.3   Seasonal Frequency of Noctilucent Clouds   89
7.4   Climatology of the Mesosphere   92

8   Other Observations   97

8.1   Introduction   97
8.2   Association with Hydroxyl Airglow Emission   99
8.3   Association with Aurora and Planetary Magnetic Activity . . .   100
8.4   Lunar Effects   104
8.5   Lidar Observations   104
8.6   Artificial Noctilucent Clouds   105
8.7   Abnormal Observations   106

9   Environment of Noctilucent Clouds   108

9.1   Introduction   108
9.2   Atmospheric in Temperature   108
9.3   D-Region   110
9.4   Dust   112
9.5   Water Vapour in the Mesosphere   112
9.6   Radiation   114
9.7   Rates of Growth   116
9.8   Nucleation of Ice   118
9.9   Settling of Particles   120
9.10   Modelling Noctilucent Clouds by Numerical Simulation   122

10   The Nature of Noctilucent Clouds   125

10.1   Introduction   125
10.2   Formation in Noctilucent Clouds   125
10.3   Growth of Noctilucent Cloud Particles   127

10.4   Evaporation of Noctilucent Cloud Particles   127
10.5   The Relationship Between Polar Mesospheric Clouds and
        Noctilucent Clouds   128
10.6   Summary   129

11   Bibliography   131

A)   Before 1900   131
B)   1900–1950   132
C)   Bibliography since 1950   134

Appendix 1: Atmospheric Refraction   149

Appendix 2: Atmospheric Transmission Along Grazing Pays   154

Subject Index   157

Name Index   161

1 Noctilucent Clouds

1.1 Introduction

Noctilucent clouds are immediately recognizable, even when being seen for the first time. The name suggests it all: they *are* night-shining clouds. From mid-latitudes ($\phi > 50°$), they can be seen during the summer in the twilight arch which moves around the north (or south, in the southern hemisphere) horizon as the night progresses. In form much like cirrostratus clouds, they are usually silvery-white or pale blue in colour and they stand out clearly behind the darker twilight sky. Ordinary (i.e. tropospheric) clouds are dark silhouettes under these conditions; noctilucent clouds shine. The reason for this is that noctilucent clouds are very high in the atmosphere and remain in sunlight long after the Sun has set at ground level.

Noctilucent clouds were first noted during the summer of 1884, some months after the volcanic explosion which destroyed Krakatoa Island in 1883. From the first, they were recognized as being quite different in character from clouds lying low in the atmosphere. It is slightly puzzling why the clouds were not written about in the years before 1885 (Gadsden 1983, 1984; Schröder 1975). Current views on the cause and development of noctilucent clouds suggest that the Krakatoa explosion had no direct effect in producing the noctilucent clouds; the event is significant only in having led observers to take careful note of the twilight sky and what was to be seen in it (Schröder 1975; Austin 1983).

Noctilucent clouds have an appearance that impresses an observer immediately and there are published descriptions that express almost astonishment at the way they shine in the sky. Certainly, astronomers and aurora-watchers are eager to contribute reports and simple measurements to the corpus of noctilucent cloud research. As it happens, one of the most extensive and brightest displays of noctilucent clouds in recent years occurred 100 years ago almost to the night after Smyth's enthusiastic report (Smyth 1886). Gavine (1987) received over 30 individual reports of the display over NW Europe on July 23/24 1986. Clearly, noctilucent clouds have lost none of their attraction in 100 years.

1.2 How, When and Where Noctilucent Clouds Are Seen

In anticipation of proper justification of the statements in later chapters, reference has been made above to "summer ... twilight", to "mid-latitudes ($\phi > 50°$)" and to "very high in the atmosphere". We shall see that noctilucent clouds are essentially a polar phenomenon and that the clouds are blown away from polar regions to disappear at mid-latitudes.

At noctilucent cloud level, the air pressure is a few millionths of that at sea level. The upper atmosphere is also very dry, with water molecules present only as a few in the million of the molecules making up the surrounding air. The partial pressure of water vapour at these heights is perhaps 10 picobars. The actual amount of water available for cloud formation in the upper atmosphere is therefore minute. Clearly, for clouds not only to exist there, but to be dense enough to be seen, there must be unusually cold conditions. This is confirmed by direct (rocket-borne, in situ) measurements of temperature.

At the height of noctilucent clouds, the atmosphere is relatively transparent to solar radiation and the local air temperature is set principally by mixing, by radiation from the air itself and from the Earth's surface and the lower atmosphere, and by bulk movement in the vertical direction. The global circulation of the upper atmosphere, largely the result of solar heating in the stratospheric ozone layer well below the noctilucent cloud region, imposes *upward* (cooling) movement of the air over the *summer polar regions*, with *downward* (warming) movement over the *winter polar regions*. Hemisphere to hemisphere flow closes the circulation; for details, the reader should consult a textbook such as that of Houghton (1977).

Consequently, the upper atmosphere shows the paradoxical behaviour of being colder in the summer than in the winter. This is the result of solar heating at these high levels being less important than the temperature changes caused by expansion or compression when air rises or falls.

Thus, there are already two inherently conflicting influences on the ability to see noctilucent clouds. If the observer goes to very high latitudes, noctilucent clouds may be difficult to observe because they are just forming and have not had the time to permit cloud particles to grow to observable size. If, on the other hand, the observer is at too low a latitude, the noctilucent clouds may have moved equatorwards, out of the regions of low temperature, and have evaporated. A compromise is called for, and observable noctilucent clouds are to be found at latitudes of approximately 60°–75° (Gadsden 1982; Schröder 1975).

The amount of water vapour in the upper atmosphere is difficult to estimate with precision. The Earth's atmosphere contains water vapour which is present at all heights through the action of diffusion upwards from sea level. The tropopause acts as a "cold trap" for water vapour: much of the upward flux of water vapour will be frozen out at or below the tropopause.

If we assume that the atmospheric pressure at the tropopause is typically 100 mb, and that the tropopause is at a temperature of 200 K (at which the saturation vapour pressure of water over ice is 1.7 μb), the relative amount of water vapour just above the tropopause will be 17 parts per million (ppm). This, the so-called *mixing ratio*, is the number of water molecules per given number of atmospheric molecules.

Water vapour in the upper atmosphere comes, therefore, from a dry, cold, region low down. An upward flux (and the negative gradient of mixing ratio, necessary to keep the flux going) is maintained through "evaporation" of hydrogen atoms from the topmost part of the atmosphere, the *exosphere*. As the water molecules diffuse up in the atmosphere, they may be dissociated (at stratospheric heights) by solar ultraviolet radiation to give hydrogen atoms and hydroxyl radicals. In the mesosphere, therefore, there is a complicated balance in the system of chemical reactions in which ozone and atomic oxygen, among others, play a prominent role.

What measurements are available to allow estimates of the mixing ratio at 80 km and above suggest that 3 ppm is probably as good a number as any to keep in mind.

Temperatures as low as 111 K (-162°C) have been measured at a few kilometres above a noctilucent cloud. At such low temperatures, water molecules will cluster together quite effectively to form ice particles although the flux of molecules to a surface is small. The number of water molecules in the atmosphere above, e.g. 80 km is approximately 7×10^{15} m^{-2}, that is a mass of 0.2 μg m^{-2}. The entire water content above 80 km could freeze out to give just one cloud particle of radius 0.042 mm for each square metre of cloud layer (or more particles, of course, inversely as the cube of their assumed radius). One does not expect, therefore, very *dense* noctilucent clouds and measurements show that their optical thickness is usually in the vicinity of 10^{-4}; noctilucent clouds are transparent and scatter less than one part in a thousand of the sunlight incident on them.

They cannot be seen from sea level during the day. Light that is scattered low in the atmosphere completely hides the tiny proportion of sunlight that is scattered from a noctilucent cloud. After sunset, the sky darkens and the sky brightness is down by a factor of several hundred at the end of civil twilight (which is defined as being the time when the centre of the solar disc, with no allowance for refraction, is at a zenith distance of 96°, that is, 6° below the horizon). At this time, noctilucent clouds can be distinguished, albeit with difficulty, right across the observer's sky, from the bright twilight arch in the direction of the set Sun to the opposite part of the sky where the so-called Earth shadow is rising (Minnaert 1954).

The invisibility of noctilucent clouds during daytime and early twilight imposes another constraint, and a strict one, on successful observation. Noctilucent clouds occur only in summer (but this statement is not quite true) and only at high to middle latitudes. At latitudes higher than 61°, civil twilight lasts all night in the middle of summer and the sky does not become dark enough to permit noctilucent clouds to be seen. At high latitudes, therefore, noctilucent clouds are not to be seen from sea level except at the very beginning and at the very end of the "observing season", which is usually taken to be the months of May, June, July and August in the northern hemisphere, and the months of November, December, January and February in the southern hemisphere. Statistics of visual observations assembled by Fogle (1965; Schröder 1966c) show that 57°N is essentially the best latitude for seeing noctilucent clouds, which occur typically some 3°-5° further north. This statement is true for both hemispheres, because there is poor summer weather at latitudes of 55°-60° in the southern hemisphere.

1.3 Amateur Observations

Unless a professional scientist is living and working at a site between 55° and
60°N latitude, it is difficult for her (or him) to justify a programme of noctilucent
cloud observation. Noctilucent clouds occur perhaps on one night in five during
3 months of the year; the sky is likely to be cloudy on at least 4 nights each week
unless the location is singularly favoured by fine weather. Thus, from 365 nights
per year, there will be successful noctilucent cloud observations on perhaps 10
or 11 nights. Maybe half of these nights will be nights on which a noctilucent
cloud is barely to be seen: detectable, but not measurable. The sky is really dark
enough to show a noctilucent cloud with good contrast perhaps for 4 h in any one
night. The annual programme for observation might result, therefore, in only
20–25 h of data taking. This is not an attractive prospect for an employer or a
funding agency (even though it is far better than the prospects for solar eclipse
expeditions)!

As a result, the great majority of noctilucent cloud data by ground level
observation comes from amateur observers. The data from national groups of
observers have been published in a variety of places. The observations from the
United Kingdom are published annually in a series in the *Meteorological
Magazine*, and continue a series stretching from 1964 to the present day (Paton
1965, 1966, 1967, 1968, 1969, 1970, 1971, 1972; McIntosh and Hallissey 1974,
1975, 1976, 1977, 1978, 1979, 1980, 1981, 1982, 1983; Gavine 1984, 1985, 1986).
A more limited sequence from Poland has been published by Kosibowa and her
colleagues (Kosibowa and Pyla 1973, 1980; Kosibowa et al., 1975, 1976, 1978;
for Germany see Schröder 1966c, 1967c, e). Fast (1972) has published a
catalogue of observations gathered worldwide over the years 1885 to 1970. Other
amateur observations, from Finland and Denmark in particular, are published
sporadically and often in magazines intended to be read by amateur astronomers. Their reports, and others from northern Europe, are now appearing more
frequently in the annual reports published by Gavine (loc cit).

The British observations, and these are being extended to comprehend
observations from the rest of northern Europe, are collated under the aegis of the
British Astronomical Association (Aurora Section). A small pamphlet is
available containing advice on recording observations and on what to measure
(azimuth and elevation of features, angular extent, etc.), on how to estimate
brightness, how to note structure (veil, bands, waves or whirls) and to check
whether an aurora is present or not at the same time as the noctilucent clouds.
Forms are available for providing a standard format of reports.

An analysis of some 20 years' amateur observations has been published
(Gadsden 1985) and some of the data described in later chapters comes from
these amateur observations. Amateur measurements often involve the simplest
of equipment; for example, an alidade made from a protractor and a plumbline
will give quite precise measurements of elevation angles.

Many of the observers take pictures of noctilucent clouds and experience
shows that these are both easy to obtain and full of information. The most

impressive pictures come from using the so-called large format cameras, with a 6 × 6 cm or larger picture area. Nevertheless, 35-mm cameras, typically with 40- or 50-mm focal length lenses, can make very satisfactory records. The correct exposure time, with 200 ASA (24°DIN) film and a lens focal ratio of 2.8, will vary from 20 s at midnight to 2 s close to civil twilight. Colour film, either slide or negative (print), is almost invariably used. The *time* at which the photograph was taken should be noted with care and precision. If the photograph is exposed at 0, 15, 30 or 45 min past any hour (and timed to be within a few seconds of these times) then it is possible that the photograph can be subsequently matched in time with one taken from some distance away by another observer. Precise heights of the noctilucent cloud can be found from such pairs of photographs by triangulation, using the star background to calculate azimuths and elevation angles.

Movie or cine cameras are used in time-lapse photography; one exposure every 3 or 6 s gives an appropriate and convenient speeding up when the film is viewed at 24 or 16 frames s^{-1}. The exposure time can fill the interval between frames if a lens of focal ratio 2 or less is used.

Sometimes video cameras have been used; a silicon cathode and lens of focal ratio 2.8 or 4 will give a satisfactory signal for tape recording.

1.4 Cloud Types

The standard reference for description of cloud forms is the "International Noctilucent Cloud Observation Manual" (WMO 1970). This gives the following taxonomy of noctilucent clouds:

Type I — Veils. Very tenuous, lack well-defined structure and are often present as a background to other forms.

Type II — Bands. Long streaks, often occurring in groups arranged roughly parallel to each other or interwoven at small angles (Fig. 1.1).

Type III — Billows. Closely spaced, roughly parallel, short streaks (Fig. 1.2). Sometimes lie across the long bands, giving the appearance of a comb or feather.

Type IV — Whirls. Partial or complete rings of cloud with dark centres (Fig. 1.3).

Type V — Amorphous. Similar to veils in that they have no well-defined structure, but they are brighter and more readily visible than veils.

One or more of the individual types can occur on one evening, and even simultaneously in different parts of the sky. One of the most spectacular displays of recent years was seen from northern Europe on the night of July 23/24 1986. From evening twilight to dawn, noctilucent clouds were seen filling the twilight arch. The vividness of the display was increased by the lower atmosphere being particularly clear at this time; for an appreciable length of time during the middle of the night, the upper edge of the clouds was deep red. This was caused by the sunlight incident on the clouds having been able to penetrate the

Fig. 1.1. Noctilucent cloud *bands*, photographed on July 4, 1975 at 2351 UT from Aberdeen, Scotland

Fig. 1.3. Noctilucent cloud *whirls*, photographed on June 27, 1972 at 0105 UT from Aberdeen, Scotland

atmosphere to unusually low levels at the sunset line (see Sect. 3.4), many hundreds of kilometres to the north of the observers.

A series of photographs from the display on this night is shown in Figs. 1.4 to 1.7, where they are presented in chronological sequence. The display was noticeable in the northwest a couple of hours after sunset (Fig. 1.4) with irregular billows lying across a system of fainter bands.

Almost an hour later (Fig. 1.5), the noctilucent clouds are confined to a much lower twilight arch, in the north, and show a brilliant whirl of cloud lying within billows and bands. The explanation of such a structure in naturally-occurring noctilucent clouds is not simple. The picture is reminiscent of the appearance of a wind-distorted rocket trail: perhaps a meteor has nucleated the noctilucent cloud some hours before, to start off what could have been, at first, a straight cylindrical cloud form.

◄——

Fig. 1.2. Noctilucent cloud *billows*, photographed on June 29, 1975 at 0035 UT from Aberdeen, Scotland. Note the greater contrast of the image reflected in the window at the *top* of the picture. This is caused by polarization in the light from the noctilucent cloud

Fig. 1.4. Noctilucent cloud photographed from Aberdeen, Scotland, on July 23/24, 1986 at 2237 UT: direction northwest

Fig. 1.5. Noctilucent cloud photographed from Aberdeen, Scotland, on July 23/24, 1986 at 2324 UT: direction north

Fig. 1.6. Noctilucent cloud photographed from Aberdeen, Scotland, on July 23/24, 1986 at 0031 UT: direction north

Fig. 1.7. Noctilucent cloud photographed from Aberdeen, Scotland, on July 23/24, 1986 at 0212 UT: direction northeast

In Fig. 1.5, the original colour slide shows a red upper edge to the clouds and, after the lapse of another hour, the display has filled the twilight arch and showed the red edge very well (Fig. 1.6).

Finally, the display (Fig. 1.7) fades in the northeast into the brightening dawn sky. On this occasion, the noctilucent cloud shows a definite southern boundary with a rippled edge, billows with wavelike modulations running along the billows in a direction perpendicular to the direction of the billow waves themselves.

1.5 Structure of the Upper Atmosphere

The Earth's atmosphere, although in movement in both the horizontal and vertical directions, is in essence always close to being in hydrostatic equilibrium. That is, atmospheric pressure (p) decreases with increasing height (z) at a rate close to that given for perfect hydrostatic equilibrium:

$$dp/dz = -\rho g, \qquad (1.1)$$

where ρ is the atmospheric density at height z and g is the local acceleration due to gravity. As Chapman (1931) pointed out in his classic paper on the upper atmosphere, this approximation is good over layers which are quite extended in height. The equation may be integrated to give

$$p = p_o \exp\{(z_o - z)/H\}, \qquad (1.2)$$

with p_o being the atmospheric pressure at height z_o and H (the "scale height") written for the quantity (kT/mg). In this expression, T is the temperature in Kelvin of the atmosphere (assumed constant over the layer), k is Boltzmann's constant, g is the acceleration due to gravity and m is the characteristic or mean molecular mass in the layer.

The change of temperature with height above sea level is shown schematically in Fig. 1.8. Sunlight, passing down through the atmosphere, is principally absorbed on reaching the surface, whether ground or water. The surface is thus a major source of heat for the atmosphere. Conduction and convection in the lower atmosphere is sustained by the lapse of temperature (the negative gradient with height) extending from the ground up to the tropopause. In the stratosphere, absorption of solar ultraviolet radiation by ozone provides a second source of heat and there is a maximum temperature at the top of the stratosphere. (Although most solar radiation taken up by ozone is absorbed low in the stratosphere, just above the peak concentration of ozone at about 20 km, the heat capacity of the atmosphere decreases rapidly with increasing height. Consequently, the temperature of the atmosphere shows a maximum well above the level of maximum absorption of energy.)

At greater heights still (above approximately 95 km), solar ultraviolet radiation and X-radiation is absorbed; solar energy is deposited in the at-

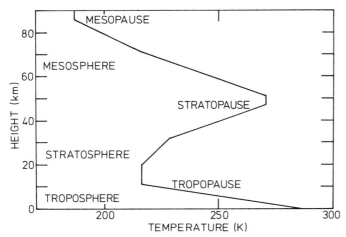

Fig. 1.8. A sketch of the variation with height of temperature of the Earth's atmosphere, in which the model atmosphere numbers listed in the US Standard Atmosphere 1976 (NOAA 1976) have been plotted. The principal regions of the atmosphere, divided on the basis of temperature variation, are labelled and given the names in common and widely accepted use

mosphere through the processes of photodissociation and photoionization. At great heights, above the range shown in Fig. 1.8, the atmospheric temperature rises to several hundred or a few thousand degrees. Between the hot, topmost part of the atmosphere (the thermosphere, and the magnetosphere higher still) and the warm stratosphere beneath, there is a deep minimum in temperature, the mesopause.

The mesopause occurs, by definition, at the top of the mesosphere and at the bottom of the thermosphere. Noctilucent clouds appear always in the vicinity of the mesopause. Consideration of the physics of nucleation and growth, with settling of the cloud particles, suggests that noctilucent clouds can be expected to be a little *below* the mesopause. Recent measurements of noctilucent clouds from rocket-borne sensors have shown that on occasion the mesopause is split; there may be several minima in temperature which are separated by a few kilometres in height.

The region containing the mesopause is not easily accessible to in situ measurements. It is at a height far above the ceiling for balloons (30–40 km) and too low for a satellite to survive one orbit before re-entry. The region is within reach of small (single-stage) sounding rockets but the flight of a rocket provides just two spot measurements of the mesopause, one on the way up, and one on the way down. The measurements are separated by a minute or two in time and are made at sites up to a few kilometres apart. Observations of noctilucent clouds, on the other hand, provide a way of watching the behaviour of the atmosphere just below the mesopause for a period of hours, allowing the observer to estimate wind speeds, to study wavelike movements and to measure quite precisely the height at which the atmosphere is unsaturated with respect to an ice particle falling into surroundings having ever-increasing temperature. The simple statistics of occurrence of noctilucent clouds indicate the response of the upper mesosphere and lower thermosphere to solar activity. And it may well be that noctilucent clouds can be used to follow the movement of distinct air masses from the pole during the summer circulation pattern. Certainly one of us (Schröder 1968a,b,d, 1974) has discussed at length how the start of the noctilucent cloud "season" and its end seem either to show or to be controlled by the abrupt arrival and equally sudden end of summer in the upper atmosphere.

2 History

2.1 Introduction

Long before the actual discovery of noctilucent clouds in 1885, reports had been published describing several varieties of different phenomena seen visually in the upper atmosphere. The existence of the aurora borealis has been known since classical times but no physical explanation of the phenomenon was published before the 18th century (Schröder 1984). Many papers dealt with phenomena such as halos, abnormal colours in the aurora, striking cloud formations and extensive and intense meteor showers. It is rather surprising that in the early literature no reports seem to describe observations of noctilucent clouds (Archenhold 1928; Förster 1906, 1908, 1911; Hoffmeister 1951; Jesse 1884, 1886; Schröder 1966c; Störmer 1925; Peters 1894).

An early account (Maignon 1648) does not refer to noctilucent clouds (Gadsden 1985) despite attributions often made (Bronshten and Grishin 1970). Some of the early descriptions of aurora borealis (de Mairan 1733) probably refer to enhanced airglow rather than to noctilucent clouds. He reported that for several years and over periods of 15 successive nights, and even for entire months, the phenomenon was vaguely spread in the sky.

In another place, de Mairan pointed out that those phenomena which he has called "horizontales" (because they spread their light at a restricted height not only towards the north but sometimes all around the horizon) should be classified as quiet aurorae (cf. for the history of geophysics Ertel 1953).

In 1783, extensive air pollution and corresponding changes in the colour of the sky were observed in Europe. These were described by Verdeil (1783) and have been discussed by Hersé (1988). From these reports one can infer only that there was a generally enhanced radiance of the twilight sky but not that there were noctilucent clouds. Deluce (1787) reported having seen a luminous mottle covering a zone several degrees wide extending from the orient to the occident.

However, this report, and a later one by Arago (1838), refer to a general brightening of the night sky, and observation of noctilucent clouds is not to be inferred. Schmidt (1869) noted a general, diffuse illumination of the night sky occurring in the course of observations of meteors in 1853, 1855 and 1861.

Marron y Miranda and Manuel (1899) wrote about a pseudo-aurora borealis being seen at times of the great Leonid meteor showers, with examples for the years 1789, 1863 and 1866. General radiance of the night sky has been noted also at the time of other meteor streams (Andromedids and Bielids);

many reports can be found in the records of the aurora section of the British Astronomical Association, including observations of non-polar aurorae (i.e. aurorae without concurrent magnetic activity). In Germany the Heidelberg astronomer M. Wolf devoted much attention to these general sky illuminations. Recently, Baggaley re-opened discussion of these events (Baggaley 1977) and Gadsden (1980) has shown that the commentators are, in general, mistaken and that there is either magnetic activity at the time (and therefore the strong possibility of aurora) or that the reports of the glows are so indefinite that no firm conclusions may be drawn. The modern meteor storm (that of the Leonids in 1966) made no detectable effect on the night airglow either before, during or after the storm (Gadsden loc cit; Schröder 1962, 1966c).

However, nowhere in all these reports, nor in the twilight research discussed by Kießling (1885, 1888), are noctilucent clouds described in a recognizable way.

2.2 The Discovery of the "Shining Night-Clouds"

In 1883, during the First International Polar Year (1882–1883), disastrous volcanic eruptions occurred on Krakatoa, an island near the coast of Java. On August 26, the volcanic activity caused noise loud enough to be heard some 150 km away. The following morning, some two-thirds of the island (20 km²) was destroyed and sank into the sea. The shock waves from this explosion travelled around the Earth several times and were registered on barographs at many meteorological stations. Calculations of the speed of the waves gave estimates varying from 302–322 m s⁻¹ (see also Verbeek 1885; Symmons 1888).

The eruption also resulted in atmospheric effects and, according to Kießling (1888) and Schröder (1975), four periods of development can be traced:

1. In the first period, from the end of August to the end of September, a belt of volcanic debris developed around the Earth, the belt closing the complete equatorial circle in about 12–13 days. From this an equatorial high-altitude air current from east to west with a speed of 36 m s⁻¹ was inferred. The dust gradually dispersed from the equatorial region and formed two latitudinal zones across the northern and southern hemispheres.

2. In the second period, from the beginning of October to the first half of November, the dust clouds moved to higher geographical latitudes. The most striking abnormalities were observed between 20° and 30° latitude (both N and S). During this time, intense twilight colours occurred over large areas of the Indian Ocean, over the northwest coast of Africa and elsewhere. During the first days of November, strong twilight colours were reported from England and Denmark.

3. The third period, from the middle of November onwards, brought a sudden increase of optical disturbances over the whole northern hemisphere. On November 23, the phenomena were widely reported in the USA and Canada, and shortly afterwards similar observations were reported from

Iceland. Abnormally coloured twilight skies were observed in the whole of Western Europe and these rapidly spread to as far as East Asia.

4. The final stages of the phenomena lasted until about the summer of 1886. Concerning this period, a letter by Backhouse (1886c) sent to the Krakatoa Committee of the Royal Society of London is of interest:

I first observed here [Sunderland, England] a strange sunset and sunrise on the evening of the 25 November but I understand that an unusually fine sunset was seen here on the 24th.
My first distinct observation of the pink circle around the sun by day was on the 26 November, but it had doubtless existed before then, as I had frequently observed pink colour near the sun for at least a fortnight, perhaps even a month, previously although I had not recognized its circular form.
The pink semi-circle opposite the sun near sunrise and sunset was first noted on the 27 November. Both these circles have been constantly visible since that time up to the present, when circumstances have been favourable.
The wisps of dust or ice were at first very small, but have on the whole been gradually increasing in size, and becoming more indefinite. The direction in which their length lies varies, and I have made several observations on this point.
I have found hexagonal crystals, I think not quite equiangular as well as others nearly square, in dirt collected off the windows on the 25 Dec., left there probably by the rain-storm of the 11–12th Dec. the windows not having been cleaned in the meantime.
As regards other phenomena, there has been a remarkable absence of aurora of even moderate brightness since Oct. 5, while thunder and lightning have been unusually frequent. I do not know whether this is connected with the volcanic eruptions.

This letter by Backhouse may find a place with reports which appeared during the period before the actual discovery of the noctilucent clouds. It is certain that at the times of coloured twilight appearances of 1883/1884, no noctilucent clouds were discovered. Various reports also exist which could be interpreted as noctilucent clouds, but this will always remain uncertain (Pernter 1889; Schröder 1975; Gadsden 1985).

The first mention of noctilucent clouds in the popular scientific magazine Nature (London) was made by Leslie (1885), who wrote:

Ever since the sunsets of 1883 and last year there has been at times an abnormal glare both before and after sundown. But I have seen nothing in the way of twilight effect so strange as that of Monday evening, the 6th, when about 10 p.m. a sea of luminous silvery white clouds lay above a belt of ordinary clear twilight sky, which was rather low in tone and colour. These clouds were wave-like in form, and evidently at a great elevation, and though they must have received their light from the sun, it was not easy to think so, as upon the dark sky they looked brighter and paler than clouds under a full moon. A friend who was with me aptly compared the light on these clouds to that which shines from white phosphor paint. This effect lasted for some time after 10 p.m. and extended from west to north, the lower edge of the clouds, which was sharply defined, was about 12 degrees above the horizon.

Robt. C. Leslie

Noctilucent clouds were first seen by Backhouse in Germany in the summer of 1885 when he saw them on 8 June (Backhouse 1885a,b). At about the same time, they were described by Teraskii in Russia (Teraskii 1890) as well as by Jesse (Fig. 2.1) in Germany (Jesse 1885, 1886a, 1890a-b; Archenhold 1894). If, however, by "discovery" is meant more than just "seeing", the history of the discovery of the noctilucent clouds must be closely linked to the work of O. Jesse (1838–1901).

Fig. 2.1. O. Jesse (1838–1901), founder of the first research programme of upper atmosphere physics

Jesse, and his co-worker Stolze, took the first photographs of noctilucent clouds in 1887 (Fig. 2.2). A few months later, the Berlin Academy of Science funded Jesse's research. In subsequent years, 1889–1896, a special photographic station was established at Berlin-Grunewald, headed by F.S. Archenhold (1861–1939), who also conducted photographic astronomical studies of galactic nebulae, meteors, etc. (Fig. 2.3 Archenhold 1894, 1928). Starting in 1889 the stations operated with cameras of f/3.0 and f/3.5. At Berlin-Steglitz, Jesse established a special research post for upper atmospheric studies, an observatory for noctilucent clouds (Fig. 2.5).

From 1887 onwards, photographic determination of the altitude of the noctilucent clouds was undertaken in the Berlin Observatory (Director: W. Förster, 1832–1921; Fig. 2.4). The first experiments showed that these clouds occurred at very great altitudes in the atmosphere. From the first, photographic results were surprising in that the altitude of the noctilucent clouds was found to be 82 km; an altitude at which the occurrence of cloud formations was not considered possible. This result showed that the atmosphere extended to considerably greater heights than often at that time assumed. The observations thus provided observational proof of a large extension of the atmosphere upwards.

From about 1889 onwards the Berlin Atmospheric Programme of Observations commenced. Several stations (Berlin-Steglitz, Rathenow, Nauen, Potsdam and Warnemünde) linked by telephone carried out parallactic measurements and simultaneous photography of noctilucent clouds. Over the years, several 1000 photographs were obtained and analyzed (Jesse 1890a, 1896).

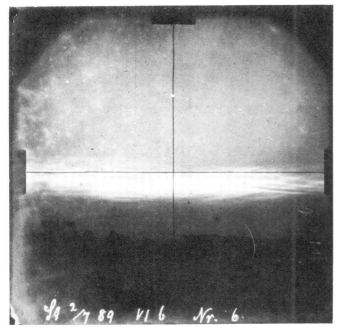

Fig. 2.2. Noctilucent cloud photography taken on July 2, 1889 by Jesse

Fig. 2.3. Friedrich S. Archenhold (1861–1939), co-worker in the Berlin Atmospheric research programme

Fig. 2.4. Wilhelm Förster (1832–1921), Director of the Berlin Observatory and sponsor of the Berlin Atmospheric Research Programme

The programme included the following phases:

1. Systematic parallactic photographs of noctilucent clouds;
2. Visual observations of noctilucent clouds to ascertain diurnal and seasonal variations;
3. Height determinations by photographic methods;
4. Spectroscopic and photometric studies;
5. Development of a general theory of noctilucent clouds;
6. Global observations of these phenomena.

2.3 Measurements of Noctilucent Clouds

From extensive altitude determinations, Jesse gave the average altitude of the clouds as 82.08 ± 0.1 km. Jesse also concluded that noctilucent clouds can extend over different heights at different times. As well as the earliest height measurements, estimates were made of the speed of movement of structure in noctilucent clouds. For the E-W direction speeds of about 109 m s^{-1} were found and for the N-S direction about 53 m s^{-1}. Thus, from these Berlin photographs (Fig. 2.5) not only the altitude but also the speed of movements in noctilucent clouds were derived (Jesse 1887a, 1890b, 1891).

While the characteristics of the noctilucent clouds were recognized correctly, the derivation of a theory for their formation was not possible. Jesse tried to find a theoretical explanation, but the decisive obstacle was the fact that sufficient data on the high atmosphere were not at that time available.

Fig. 2.5. The noctilucent cloud observatory ("Wolkenwarte") of Otto Jesse in Berlin

2.4 The Middle Period of Noctilucent Cloud Research

In the period between 1885 and 1900, several authors surmised that a relation between the noctilucent clouds and the Krakatoa volcanic eruptions might exist. Noctilucent clouds have also appeared after other volcanic eruptions (those of Katmai and in Chile in 1921). It was suggested that there was a causal relationship, whereby the eruptions, which blew particles into the upper atmosphere, caused an increase in the frequency of occurrence of noctilucent clouds. Wegener (1911) suggested that the noctilucent clouds were ice clouds, similar to cirrus clouds, with the material coming from the Krakatoa eruption. Jardetzky (1926) rejected the volcanic origin of noctilucent clouds in favour of a simple ice crystal model.

From 1880 onwards, Soviet researchers have participated in the observation of noctilucent clouds (Bronshten 1970; Chvostikov 1952; Vasil'ev and Fast 1972). The first systematic observations from the USSR of noctilucent clouds were started in 1936. After 1950 the situation changed considerably, when the USSR Academy of Sciences extended considerable effort into support of upper atmospheric research, which naturally included noctilucent clouds. I. A. Chvostikov initiated the noctilucent cloud research and during the years that followed, many detailed studies on noctilucent clouds and upper atmospheric physics have been published. The USSR programme was conducted at roughly 250 stations evenly distributed over the USSR north of 45°N, 220 of them belonging to the network of the hydrometeorological service. Other observatories were part of the All-Union Astronomical and Geodetic Society (VAGO).

During the Norwegian programme of auroral research, initiated by C. Störmer (Figs. 2.6–2.7), attention was paid to the possible observation of noctilucent clouds (Störmer 1925, 1933). Humphreys (1933) discussed early observations from the USA. Vestine (1934) published a summary of all observations known to him at that time and discussed the statistics of occurrence, movements, etc. (see Millman 1959; Currie 1963).

Research and detailed comparisons of data on noctilucent clouds with aerological data have been carried out in Canada under the guidance of A. Christie (1966). Systematic observation in North America was begun in 1962, with the co-operation of the US Weather Bureau and the Meteorological Service of Canada. The observation programme was greatly expanded during 1962–1964 to include approximately 100 stations in Alaska and Canada. Several of the meteorological stations in Greenland and Iceland and the crews of some commercial aircraft flying at high latitudes during twilight also participated in the programme.

In the UK, noctilucent clouds have been observed since about 1939 by J. Paton (Fig. 2.8) who established an important network in the following years

Fig. 2.6. Henrik Mohn (1835–1916), Norwegian meteorologist, pioneer in the research of the so-called "mother of pearl-clouds"

Fig. 2.7. Carl Störmer (1864–1957), founder of auroral research and researcher of nacreous clouds and noctilucent clouds over Norway

Fig. 2.8. James Paton (1903–1973), Scotish meteorologist, pioneer in the modern phase of noctilucent cloud research

(Paton 1964). In Germany, systematic observations were begun in 1957 by Schröder and a summary of the methods and objective of this work is given in Schröder (1966c, 1967c, e; cf. Bernhardt 1982).

In the International Geophysical Year (1957–1958), the noctilucent cloud programme was discussed at scientific meetings and supported by several international organizations. A period of global noctilucent cloud research had finally begun. With the truly global nature of the programme and standardized synoptic observations, attention could now be turned to comparisons between the northern and southern hemispheres, to studying the influence of geomagnetic and solar activity upon the frequency of occurrence of noctilucent clouds, and to analyzing the influence of thermobaric processes in the troposhere and stratosphere (Förster 1906, 1911b; Malzev 1929; Chvostikov 1966a).

One of the most recent highlights in noctilucent cloud studies has been the discovery of *polar mesospheric clouds* (see Sect. 4.5). The early satellite observations (Donahue et al. 1972) of a scattering layer at high latitudes during the summer, in either hemisphere, have been extended and greatly increased in scope by the Solar Mesosphere Explorer satellite observations reported by Thomas and colleagues (for example, Thomas and McKay 1986, and work in progress by Thomas, Olivero and Jensen). Currently, there is a series of rocket-borne in situ studies of the high-latitude summertime upper atmosphere under the embracing title of the Cold Arctic Mesosphere Program (CAMP). Table 2.1 lists developments in noctilucent cloud research in chronological order (cf. McKay 1982; Schröder 1975).

Table 2.1. Milestones in noctilucent cloud research

Year	Event
1882–1883	First International Polar Year
1883	Krakatoa eruption
1883–1885	Worldwide increased twilight phenomena
1884	Earliest (single) report of luminous night clouds
1885	First observation of noctilucent clouds by Backhouse in Germany
1885–1896	Jesse's photographic programme at the Berlin Observatory
1908	Tunguska-event
1932–1933	Second International Polar Year
1957	Ground-based optical studies
1957–1958	International Geophysical Year
1959	International Geophysical Cooperation
1963	Mesopause rocket research
1968	Rocket temperature measurements
1973	Presence of ion clusters
1978	Large $H_2(H_2O)$ clusters at 91-km height
1981/1982	Solar Mesosphere Explorer; Cold Arctic Mesosphere Programme

3 Observations from Ground Level

3.1 Introduction

Noctilucent clouds are seen because they scatter sunlight. To understand and to interpret the observations, it is necessary to consider in some detail what is involved in this scattering. It has become clear that noctilucent clouds are very thin and very sparse assemblies of scatterers. The cloud particles, whatever they may be, scatter light as individuals not as an assembly. Furthermore, there is a negligible amount of rescattering in the cloud: having been once scattered, sunlight emerges from the cloud without further scattering.

To arrive at the level of noctilucent clouds which are being seen during twilight, the sunlight that shines on the clouds will have passed through the lower layers of the atmosphere. Absorption and scattering take place along the path and these have to be taken into account and allowance made. There is also the easier problem of allowing for absorption by the part of the atmosphere that lies between the observer and the clouds. (Observations from ground level are frequently made perforce at small angles of elevation and the effect of atmospheric absorption on the downcoming scattered light should not be ignored.)

3.2 The Geometry of Twilight Scattering

There are rather close restraints imposed on the observation of noctilucent clouds from ground level. The tightest of these, apart from the need to have a clear sky over the observatory, results from the geometry of twilight. The clouds are seen by scattered sunlight and need to be sunlit; the observer needs a dark sky to see the illuminated clouds. There is, therefore, a narrow range of twilight when the clouds are able to be seen from the ground.

If absorption and refraction of sunlight in the lower atmosphere are ignored and if the Sun is treated as a point source, simple geometry gives the area of sky in which the clouds can be seen from any particular place. The shadow of the Earth on a layer of noctilucent cloud can be taken as lying within a circle of radius $(R + H)$, where R is the radius of the solid Earth and H is the smallest height (the screening height) to which sunlight is able to penetrate in its grazing passage through the atmosphere. This circle is centred a distance Y from the centre of the Earth along the antisolar axis (Fig. 3.1).

If an observer is at O, where the solar depression angle is β, he sees a point P on the edge of the illuminated area of noctilucent cloud at a slant range, S, given by

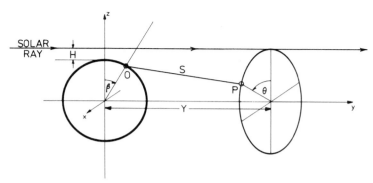

Fig. 3.1. The geometry of observation of the edge of the illuminated area of a noctilucent cloud display. The Sun is assumed to be a point source; the y-axis passes through the centres of the Sun and the Earth. An observer at O defines the plane (y,z) of the solar meridian; the Sun is at a depression angle, β, at the observer. The Earth (radius R) projects a cylindrical shadow which is increased in radius by the screening height H. The locus of P, a point on the edge of the shadow, lies at a distance $(R+h)$ from the centre of the Earth where h is the height of the noctilucent cloud. (P is shown in the diagram lying at an angle θ outside of the meridian plane)

$$S^2 = (R+h)^2 - 2YR\sin\beta + R^2 - 2R(R+H)\cos\theta\cos\beta \qquad (3.1)$$
$$\text{and } Y^2 = (R+h)^2 - (R+H)^2. \qquad (3.2)$$

In these equations, h denotes the height of the cloud layer. The point P (see Fig. 3.1) appears to the observer to be at an elevation e on a vertical with azimuth a relative to the direction of the Sun. The angles a and e are given by

$$S\sin e = Y\sin\beta - R + (R+H)\cos\theta\cos\beta \qquad (3.3)$$
$$\text{and } S^2\cos^2 a = (Y - R\sin\beta)^2 + \{R\cos\beta - (R+H)\cos\theta\}^2. \qquad (3.4)$$

Plots of the loci of P are given in Fig. 3.2 for the values h = 82 km, H = 15 km and R = 6371 km.

It is clear that if the Sun is more than 12° below the horizon, noctilucent clouds can be seen only close to the horizon in the direction of the Sun. If the Sun is less than 8° below the horizon, in principle the clouds can be seen at the zenith and beyond, in the antisolar hemisphere. In practice, however, it is found that the twilight sky is so bright at these small solar depression angles that the contrast of noctilucent clouds is very low. Dietze (1973) has considered this progressive loss of contrast as dawn approaches in some detail, as part of a wide-ranging discussion of several factors (tropospheric turbidity, dust and ozone in the atmosphere, and visual adaption) that influence the possibility of observing a noctilucent cloud (cf. Schröder 1967c). The behaviour is not dissimilar to that obtained from simple estimates.

Paton (1964) reported two occasions on which noctilucent clouds were seen at the zenith; the Sun was 7.7° below the horizon. He suggested that these observations imply a screening height of 24 or 25 km, in accord with much of his other data.

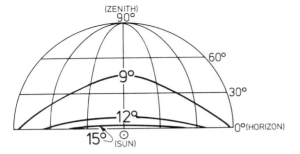

Fig. 3.2. The loci of P for solar depression angles of 9°, 12° and 15° seen projected on that half of the sunward hemisphere above the observer's horizon

The most common way of estimating the screening height is to measure the edge of the illuminated area of cloud. Plainly, interpreting these measurements to give a screening height depends on the cloud area having an edge resulting from the falloff of illumination rather than from there being a real edge to the clouds. Whether the method is appropriate on a particular occasion is the decision to be made at the time by the observer. Jesse (1896) listed a number of measurements made in 1885, 1886 and 1887 and the data are plotted in Fig. 3.3. Data from Ludlam (1957) and Paton (1964) are also given in Fig. 3.3.

If one remembers that the measured points should fall, if anywhere, to the left of a line corresponding to the calculated variation of position of the upper edge of the cloud with different solar depression angles, it would appear that a screening height of 5 km is as good a choice as any. This is much lower than Paton's own estimate of 25 km but is consistent with his data. However, in calculating the curves shown in Fig. 3.3 no allowance has been made either for the appreciable diameter of the solar disc or for the refraction of sunlight low in the atmosphere. The first effect causes noctilucent clouds to be seen a quarter of a degree of solar depression greater than would otherwise be the case. Refraction works in the same direction: for a screening height of 15 km, it is equal to 0.14°. The two effects together (for a screening height of 15 km) allow data points to fall up to 0.4° to the right of the curve in Fig. 3.3. Thus, a good choice of screening height is 15 km: 25 km is certainly too high and 5 km, for which the movement of the plotted curve is close to 1°, is likely to be too low.

This conclusion is at odds with Simmon's (1977) remark:

... Chistyakov and Teifel... found that a correction term of +30 km is required in determining NLC heights. This correction, which is now generally accepted as accurate, has been applied in obtaining the heights shown in [the] Table...

In a later paper (Taylor et al. 1984), a figure for the screening height is obtained from five separate photographs of the edge of a noctilucent cloud display taken at 10-min intervals. The photographs show cloud structure low down in the sky, lying below a clearly defined arc marking the edge of an illuminated region. A television picture recorded 195 km away simultaneously with the middle exposure of the five photographs allowed a good measurement

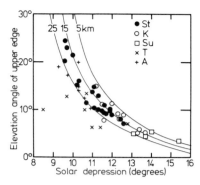

Fig. 3.3. Measured elevation angles of the upper edges of noctilucent cloud displays in the solar meridian. Data are shown from five observing sites: Steglitz (52°N), Kissingen (50°N) and Sunderland (55°N), from Jesse (1896); Torsta (63°N) from Ludlam (1957); Abernethy (56°N) from Paton (1964). The curves correspond to screening heights of 5, 15 and 25 km for a cloud height of 82 km. A point source Sun and zero atmospheric refraction have been used to calculate the curves

(82 ± 1 km) of the height of the cloud structures. This height was applied to five separate measurements of the arc elevation and a plot of the derived values for screening height showed that on this occasion the screening height was 7 ± 1 km. Taylor et al. recognized that refraction of the grazing ray would be important and made full and proper allowance for it, as also for the solar disc. They also suggested that there may have been *two* layers of cloud present, one that showed structure and lay perhaps a few kilometres below a featureless sheet of cloud. This upper layer would give the *edge* of the illuminated area that they measured; consequently, they suggested that 7 km is a lower estimate (but by no more than a few kilometres) of the screening height.

It is doubtful whether the concept of a "screening height" is basically meaningful. Taylor et al. (1984) suggested that the better dynamic range of modern photographic emulsions accounted for the lower screening height which they deduced, compared with the earlier (largely 19th century) estimates. However, it is the *spectral sensitivity* and speed of emulsions which have been extended during the 20th century, rather than the dynamic range.

The edge of a noctilucent cloud display is perceived to be where the contrast of the display behind the twilight foreground sky falls below a discriminable level. The contrast may be in radiance or in colour; the principal change in photographic technique, since the earliest days of noctilucent cloud observations, is the almost unvarying use nowadays of colour films. It may be that Simmons' colour photographs show an edge to the display that is appreciably higher in the sky than a visual observer would report. Observations of a red edge to a noctilucent cloud display have recently been analyzed by Avaste, Gadsden and Grechko (1988) and they concluded that absorption by atmospheric ozone and particularly the height of the top of the ozone layer in the stratosphere are of crucial importance.

3.3 Latitude of Observation

It can be taken as a general guide that noctilucent clouds will be seen only when the solar depression angle is between 6° and 15°. Also, the clouds are a high-latitude, summer phenomenon. This is shown clearly in the statistics given by Fogle (1964), who lists the reports of noctilucent clouds from North America for the 10 years from 1956 to 1965. In this period, the clouds were seen during 146 twilights in the May to August period (cf. Schröder 1964, 1966c).

Figure 3.4 gives a plot of the frequency of the reports for the latitude of the southernmost observer on any particular night. Fogle gives, however, no list of nights when the sky was clear and without noctilucent clouds. Thus, it is not possible to express the data as a probability of occurrence, but, two factors are apparent. First, the distribution with latitude contains the effect of distribution of population density with latitude; the two major peaks in the histogram of Fig. 3.4 correspond to Canada and Alaska. Secondly, there is a dearth of reports from high latitudes. This is because at these latitudes the summer night does not become dark enough to observe the clouds.

The broken line in Fig. 3.4 is the number of nights, at a given latitude, that the Sun goes below a depression angle of 6° during the months of May to August. (The sole observation of a noctilucent cloud beyond this line is that from a polar island station which reported a noctilucent cloud visible on September 13 and therefore outside the May to August "season".)

It should be borne in mind that these data refer to the sighting of a noctilucent cloud; the cloud itself will, in general, be at a position some degrees further to the north of the observer. Also bear in mind that these are the latitudes of the southernmost observer; it is generally, but not invariably, the case that observers to the north with clear skies will report noctilucent cloud visible to them also (cf. Paton 1964; Schröder 1967a,d).

A more meaningful representation of the data might be to plot them as a cumulative curve since nothing much can be said from ground level observations about the occurrence of clouds at latitudes poleward of 80°. The median

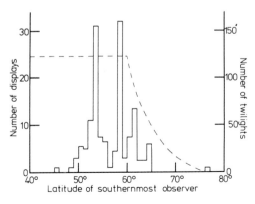

Fig. 3.4. Fogle's (1966) North American data for the appearance of noctilucent clouds in the period 1956–1965. The *broken curve* shows the number of nights in May, June, July and August when the Sun reaches a solar depression angle of at least 6°

occurrence is at 57.5° latitude; *an observer* at this latitude should see 50% of the noctilucent cloud displays which occur.

In summary there is little point in going to a latitude of more than 65° to see noctilucent clouds; the nights will not be dark enough during the summer. If the latitude of the observer is much less than 50°, the clouds will often be out of sight, beyond his horizon.

Figure 3.5 shows an outline map of the world, with the bands of latitude marked (cross-hatching) in the northern and southern hemispheres. In the southern hemisphere, most of the observation area lies over ocean. The entire Antarctic continent, except for the more northern part of the Graham Land peninsula, lies south of 65°S. In the northern hemisphere, it is clear why noctilucent cloud research is pursued so actively in the USSR, as well as Scandinavia, Germany and Scotland. In the contiguous states of the USA the latitudes are too low and in most of the state of Alaska the latitude is too high. Northern Canada is the optimal location in North America.

3.4 Absorption of Light in the Atmosphere

The sunlight that is incident on the clouds has passed obliquely through a considerable amount of the atmosphere. Because of the nature of twilight observation, this sunlight has entered the atmosphere, has plunged deeply over

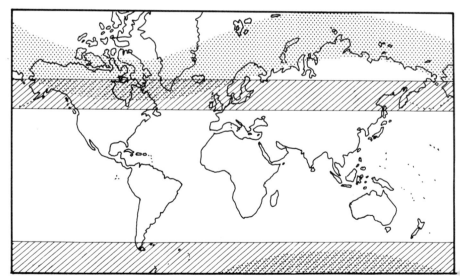

Fig.3.5. Outline map of the world. The *cross-hatched bands* of latitude extend from 50° to 65° in each hemisphere and are the "best" regions for observing noctilucent clouds. (The *dotted bands* in each hemisphere indicate zones where aurorae are likely to be visible. Reference to these zones is made in a later chapter)

the terminator and has begun to emerge again before being incident on the cloud layer. There will be significant absorption along this path.

In addition, there is the likelihood of a significant contribution to irradiance at the cloud level from light that has been scattered forward in the troposphere and stratosphere. The amount of this has not yet been calculated and its importance in the analysis of cloud observations not yet estimated. It is less of a problem for observations of noctilucent clouds from rockets (where the Sun may be well above the local horizon at cloud level) but it has become clear recently that the conditions for observation from orbiting spacecraft are in some ways no better than those for ground observation.

Rozenberg (1963, 1966) reported in detail various methods of calculating the effect of refraction, absorption (both by ozone and through molecular scattering) and the presence of aerosols on the passage of grazing rays of sunlight. The calculations are in essence simple but, on any particular occasion, the abundance of ozone near the terminator and the amount of stratospheric aerosol are not known with precision, thus the results of the calculations are generally, therefore, in the nature of general guidance to interpretation rather than precise solutions.

The effect on the colour of noctilucent clouds by the passage of sunlight through stratospheric ozone has been discussed by Gadsden (1975). Spectro-photometric measurements in the visible region show clearly an absorption band in the yellow (see Fig. 3.6); this band is the Chappuis band of ozone.

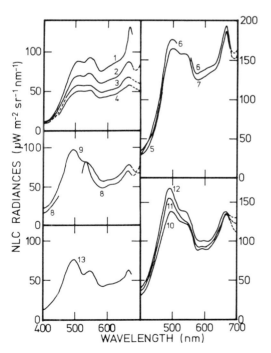

Fig. 3.6. Measurements of spectral radiance of noctilucent clouds observed from 57°N latitude on July 23, 1974. The *numbers* identify individual scans of the spectrometer and the separate *boxes* show scans on five separate areas of the clouds. Brighter features are shown on the *right*, scans *6,7* and *10,11* and *12*. In all scans, and particularly obvious in the scans on the brighter features, there is an absorption band around 600 nm

If one tries to make allowance for absorption at stratospheric levels of sunlight on its way up to the clouds, it turns out that much of the blue colour of the clouds is in fact simply the colour of the incident sunlight modified by ozone absorption. It cannot be assumed that the cloud particles are Rayleigh scatterers because the clouds are blue. Indeed, the data of Gadsden show that the clouds had a scattering cross-section that varied through the visible spectrum in a way that was very clearly not in accord with the Rayleigh approximation. However, such measurements of the spectral radiance are difficult to correct for absorption in the atmosphere of both the sunlight incident on the clouds and the light scattered from the clouds. Consequently, estimates of the characteristic size of cloud particles are given more weight when based on measurements of the degree of polarization of the scattered light than when derived from measurements of the colour of clouds (i.e. their spectral reflectance).

3.5 Height of Noctilucent Clouds

The height of noctilucent clouds has been measured directly, using photographic triangulation, on a number of occasions. The data are summarized in Table 3.1, in which the average of the heights measured on each night is listed. Jesse (1896) reported the results from eight nights on which noctilucent clouds were seen, and his summary can hardly be improved upon even to the present day, 89 years later.

von dem Jahre 1885 bis 1891 sehr nahe eine und dieselbe mittlere Höhe von 82 Kilometer gehabt haben.

The median of the observations listed in Table 3.1 is 82.9 km but this is not a serious modification of Jesse's conclusion. What is surprising is how rarely measurements of height have been done in the 9 decades that have elapsed since such measurements were first attempted. The body of data is not sufficient to permit a discussion of trends, such as any systematic change as the summer progresses, or a relationship to lunar tides in the upper atmosphere, or with geomagnetic activity at the time, and so on.

The data given in Table 3.1 are, with two exceptions, the arithmetic means of the heights measured from one or more stereo pairs of photographs taken on an evening. The measured heights commonly show a spread which is larger than the estimated uncertainty in an individual measurement. Much of this spread may be due to a systematic variation in height from place to place in the cloud. Indeed, Störmer (1933) in discussing his measured heights, stated:

A stereoscopic view of the two pictures from Oslo and Kongsberg gives the impression that the clouds were arranged in two layers of different altitude. This is in accordance with the measured heights, 83 and 84 km for the points 2 and 3 and 88, 92 and 90 km for the points 4, 5 and 6.

Witt (1962) discussed the stereophotographs which he obtained on the display visible over central Sweden on the night of Aug. 10, 1958. He used

Table 3.1. Summary of measured heights of individual displays

	Date	km
Jesse (1896)	1889 Jun 22	82.9
	Jul 2	82.5
	Jul 24	85.5
	Jul 31	88.5
	1890 Jul 6	83.0
	Jul 10	81.7
	Jul 24	82.95
	1891 Jun 25	81.4
Störmer (1933)	1932 Jul 10	81.8
	Jul 24	81.1
Störmer (1935)	1934 Jun 30	82.2
Paton (1951)	1949 Jul 10	84 to 89
Burov (1959)	1958 Jun 16	82.6
Dirikis (1962)	1959 Jul 14	83.3
Dirikis et al. (1966)	1961 Jun 30	83.1
	1964 Jun 30	82.9
Burov (1966)	1964 Aug 2	83.2
Burov (1967)	1965 (Jul 15)	(73.0 to
	(Jul 19)	96.8)
Grahn and Witt (1971)	1958 Aug 10	83.7
	1965 Aug 5	83.2
	Jul 20	82.0
	1967 Jul 16	82.2
	Aug 9	83.0
Taylor et al. (1984)	1979 Jul 10	82.0
	Median of 23 displays	= 82.9 km

Heights measured by rocket-borne photometers; in the third column the first figure given is for the bottom of the scattering layer, the second figure for the top

	Date	km
Rössler (1972)	1970 Aug 10	82.5
Tozer and Beeson (1974)	1971 Jul 31	81.9 to 83.7
	1973 Aug 1	83.2 to 85.1
Witt (1969)		82.4 to 84.4
Witt et al. (1971)	1971 Jul 31	82 to 83
	Aug 1	85 to 87
Witt et al. (1976)		85.5 to 89

precision phototheodolites (placed at his disposal by the Bofors Company) at each end of a geodetically surveyed baseline of length 51.5 km. He also had the good fortune to have almost perfectly clear conditions to make his measurements on an extensive, bright display. Two pairs of stereophotographs are reproduced in Figs. 3.7 and 3.8.

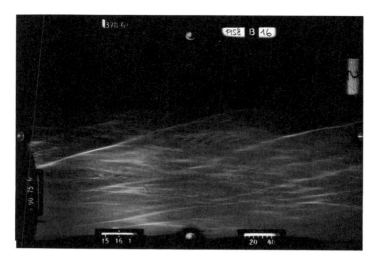

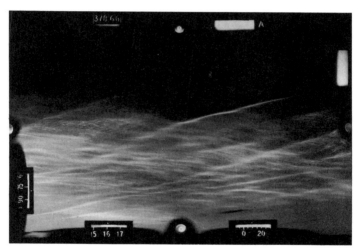

Fig. 3.7. Pair of stereogrammetric plates obtained by Prof. Georg Witt in 1958 (Reproduced here by kind permission of Prof. Witt)

In his paper, Witt is principally concerned with cloud sections and contour maps of the clouds; he gives many detailed estimates of velocity of the movement in the clouds. The range of measured heights was only 4 km; the data ranged between 81.5 and 85.5 km. In Table 3.1, the average of the individual heights listed by Grahn and Witt (1971) for this display is 83.7 km.

It should be noted that the work reported by Witt (1962) is the best that has ever been done on height measurement of noctilucent clouds and, in our opinion, the quality of the work is unlikely ever to be surpassed. This work

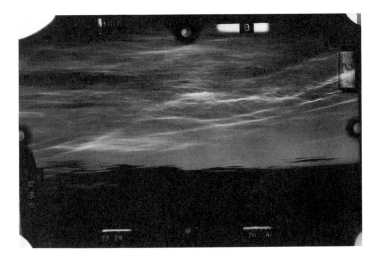

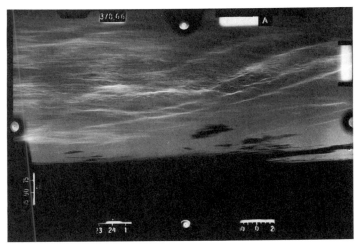

Fig. 3.8. Pair of stereogrammetric plates obtained by Prof. Georg Witt in 1958 (Reproduced here by kind permission of Prof. Witt)

should be required reading for all who attempt similar observations. It even includes a pair of anaglyph pictures, together with red and green viewing filters, to allow one to view the cloud pictures stereoscopically.

The data show that noctilucent clouds have a varying height across the field of view and also may appear at two distinct levels in the atmosphere. However, there is a need for caution in the interpretation of pairs of photographs. First, it must be remembered that the clouds are optically very thin (stars shine through them with little diminution in light), and the structure that may be obvious to the observer can result from the superposition of two separate clouds in the line of sight. Very wrong heights will be obtained from triangulation or stereogrammetry on a feature which has no real existence, and which appears at different locations in space from the two observing points. Secondly, the clouds are almost always viewed very obliquely. Once more, bright features may have no physical reality but may arise simply from seeing a thin layer of cloud with waves in it. Bright features may result from viewing in a direction in which the line of sight travels a considerable distance within the layer. The same favourable view may not be available to a second observer some distance from the first. Examples of this appear in Figs. 3.7 and 3.8 and these pairs of pictures are worth careful study in order to compare the details that appear in them.

Simmons and McIntosh (1983) have used a "statistical" method for measuring the heights of noctilucent clouds. They used, for basic data, 1680 observations made over a 17-year period. Each of the observations included the date and time of observation, the observer's latitude and longitude and the angular elevation of the edge of the cloud displays. (This was assumed to be in the solar meridian at the time of the observation.) Applying plane geometry, and assuming a screening height of 30 km, they found an average height of 83.76 km with a sample standard deviation of 15.04 km. The results lying within one standard deviation numbered 1053 (62.7%); 87 (5.2%) were beyond three standard deviations. The distribution was, therefore, really quite close to a normal distribution with an extra number (83; the excess of 5.2% over the expected 0.27%) lying in the wings of the distribution.

Little weight should be given to the quoted figure of 83.76 km. The upper edge of a display may be a true edge and not the sunset line on a cloud layer; in such a case, the measured height is an underestimate and the average of the sample distribution will also be an underestimate (Gadsden 1985). Simmons and McIntosh applied 30 km for the screening height, which we have seen is an overestimate. It is likely that there is a *fortuitous balance* of an underestimate and an overestimate to give an average height which agrees adequately with two-station measurements.

Taylor et al. (1984) described a novel and powerful method for determining the height of noctilucent clouds. The triangulation is not done directly, through measurement of parallax, but by using an iterative method which exploits the speed of computers. Photographs taken from two (or more) locations are measured to give "picture coordinates" of noctilucent cloud features and of stars appearing in the background. Each picture is then provided (though the use of the calculated positions of the stars in the pictures) with a conversion from picture coordinates to (local) azimuth and altitude (see Hapgood and Taylor 1982 for a detailed description of this method).

A height for the noctilucent clouds is assumed; for each feature in each picture the corresponding ground position, latitude and longitude is calculated. A number of assumed heights (chosen with intelligent foresight) is taken: that assumed height for which the two (or more) ground positions are coincident must be *the* height of the cloud feature.

Clearly this is a powerful and convenient method that allows the operator to judge the *quality* of the match and thus the quality of the initial data. In the example given by Taylor et al. (1984), the height of several cloud structures was found to be 82 km (and certainly not 80 km or, presumably, 84 km) in much of an area of sky some 17° (horizontally) × 16° (vertically).

Witt's method of measurement involves the full use of photogrammetric technique and equipment. The cameras are phototheodolites placed at the ends of an accurately surveyed baseline. The two camera operators remain in radio or telephonic communication throughout the display. The pairs of (glass) plates are measured on a photogrammetric plotting table by a skilled operator of the "floating-mark" technique (see, for example, Hallert 1960). However only a few lucky scientists command such facilities.

In Scotland, there is currently a programme to obtain and measure suitable pairs of photographs. The participants are amateur astronomers who use 35-mm format cameras and film for colour slides. When an observer sees a noctilucent cloud display, he bolts his camera to a simple metal bracket which tilts it up at an angle of approximately 12°; the local horizon appears at the bottom of the field of view. He can photograph an area of the display which is approximately 20° vertically and 30° horizontally. Usually the bracket has been bolted into permanent position some months before. By taking a time exposure on a clear dark night to record star trails, the observer has a record of the exact orientation of the field of view and of the lens distortion in mapping azimuth and elevation onto (x, y) coordinates in the camera frame.

The noctilucent cloud exposures are made at 0, 15, 30 and 45 min. past the hour and are timed to start exactly on the mark (to within a second). During the weeks that follow the night of the display, it becomes known whether other observers have also photographed the display. If so, then there are pairs of photographs for measurement and subsequent triangulation.

The observers' positions are measured from the national Ordinance Survey to give latitude and longitude (to within 100 m), and their heights above mean sea level. The methods described extensively by Störmer (1955) are used to calculate a *base axis*, which is the line joining two observers, and to identify the positions of the *base poles* on the celestial sphere. Corresponding points in the noctilucent cloud display, which are common to both observers' pictures, are identified in the system of spherical coordinates (ζ, ε) centred on the base axis. The relative displacement of the same point in the cloud takes place *along a great circle of ζ* (an invaluable check on the correct identification of corresponding points) and the parallactic displacement, p, is simply the difference between the two values of ε. Given p, the calculation of the geographical position of the point in the noctilucent cloud and its height above sea level is a matter of simple trigonometry.

Finally, the results of in situ measurements of noctilucent cloud heights have been included in Table 3.1. There have been several rocket flights made with photometers in an upward-looking position on board. If a rocket should happen to penetrate a noctilucent cloud, the photometer record shows a distinct and abrupt fall in radiance when plotted against time of flight. Often there is an equally distinct jump in radiance on the downward leg. The height of the rocket is usually known at all times during the flight to within a few metres so that the photometer record gives directly the heights of the bottom and top of the cloud layer.

3.6 Drift Motions

Noctilucent cloud data can provide information on the wind velocity at the mesopause level. In principle, one simply observes the apparent movement of structures in the noctilucent clouds and, assuming a height and entirely horizontal movements, the wind velocity may be calculated. Many researchers have discussed data based on this method, including Schröder (1969), Vestine (1934) and Witt (1962).

Jesse (1888, 1890b) published the first data on noctilucent cloud speeds from photographic measurements made in 1887. He found some very high speeds (up to 300 m s^{-1}) with the majority of observations lying in the range 40–177 m s^{-1}. Jesse's co-worker Archenhold (1928) published a number of measurements of noctilucent cloud data from the Berlin observations between 1885 and 1894 (cf. tables 3.2 and 3.3).

Measurements were obtained by the veteran Norwegian auroral researcher Störmer (1933, 1935). He reported an average speed of about 75 m s^{-1} (based on 47 determinations), with a sharp concentration of directions towards the SW.

Bronshten and Zatejssikov (1938) used observations from the Soviet Union to obtain an average velocity of 45 m s^{-1}, with two maxima in the distribution of speeds: a principal one at 20–40 m s^{-1} and a secondary maximum at 80–100 m s^{-1}.

Ludlam (1957), from theodolite observations, found average values of 45 m s^{-1} and with extreme values of 13–90 m s^{-1}, the directions showed a concentration towards the south (SSE to SW).

Fogle (1966) has summarized many velocity measurements and found an average cloud speed value of 40 m s^{-1} in the direction of 240° (WSW). A compilation of 67 measurements in the USSR presented by Burov (1966) and Dirikis et al. (1966) show rather similar conclusions. Burov (1966) reported also on determinations of the vertical velocity of movement of noctilucent clouds. He determined an average vertical component of 8.3 m s^{-1} (ten measurements), whereas Dirikis et al. (1966) and Dirikis and Mookins (1966) determined 11.5 m s^{-1} (32 measurements). It has been found that over a short time interval parts of the noctilucent cloud stopped their ascent and started to descend, at the same time changing the azimuth of motion. The horizontal velocity changed little.

Quiroz (1964) had earlier suggested that this behaviour is not unlike the pattern of vertical wind speeds revealed in rocket measurements. Rosenberg and Edwards (1964) suggested that both atmospheric tides and internal gravity waves may be responsible for the behaviour.

Table 3.2. Velocities. From German observations communicated by Archenhold, Hoffmeister, Jesse, Meier and Schröder

Date	m/s	Date	m/s
19.7.1885		12.7.1894	23
	74		17
22.7.1885	15		61
2.7.1889	94		36
	177		38
	125		105
	131	18.7.1951	43.6
	137	12.7.1953	84
	122		106
9.7.1889	38		121
	33		103
6.7.1890	49	21.7.1960	35
	44	7.6.1962	38
	79	27.7.1963	60
	76		57
	84	9.6.1964	38
	99		45
	69	16.7.1964	39
9.7.1890	47		49
	63	7.7.1965	51
10.7.1890	69		57
	73		67
	24	24.7.1965	40
	25		27
	28		38

Table 3.3. Noctilucent cloud speeds

Author	ms^{-1}	Range	n
Jesse (1896)	69	15–208	61
Störmer (1935)	57	44–83	8
Bronshten and Zatejssikov (1938)	45	17–135	23
Ludlam (1957)	45	13–90	17
Witt (1962)	90?	60–120	?
Fogle (1966)	98	85–110	2
Dirikis et al. (1966)	74	31–180	37
Burov (1966, 1967)	76	17–262	30
Schröder (1974)	75	20–125	27

Summarizing data is often very subjective. Some authors give ranges of direction and speed which reflect the complexity of a cloud display (Tables 3.2 and 3.3). Quiroz (1964) has produced a frequency distribution of estimated velocities (i.e. speeds and directions) using published data for observations made in the years 1885–1940 and 1954–1955. These data, reproduced here in Table 3.4, show that in 75% of the cases, noctilucent clouds move *from* directions lying between north and east, with ENE as the predominant direction. In 73% of the cases speeds were in the range of 26–100 m s^{-1}. The highest speeds were those when the clouds were seen coming from the north or northeast. The speeds seem unusually high for the atmosphere in the height range 75–85 km; maximum winds observed by the rocket-borne grenade technique are usually of the order of 20 m s^{-1} (cf. Nordberg et al. 1965b; Theon et al. 1969b).

Since the date of Quiroz' compilation, Simmons (1977) has published an analysis of photographic observations of the noctilucent cloud display of June 18–19, 1976. His deduced speeds vary from 34 to 51 m s^{-1}, rather lower than the typical values quoted above, in directions 186°–197°.

Table 3.4. Distribution of cloud velocities (after Quiroz 1964)

Direction, from		SW	W	NW	N	NE	E	SE	S	Total
				1885–1940, 1954–1955						
Number, all speeds		1	4	5	14	29	34	5	6	98

Direction	< 26	26–50	51–75	76–100	101–150	> 150
		Number of cases, according to speed				
SW	1	1				
W						
NW	1	1		2		
N		4	1	2	1	1
NE	1	1	4	2	1	1
E	1	4	2	2		
SE		1		1	2	
S	1	1		1		
Totals:	5	13	7	10	4	2
Total (%):	12	32	17	24	10	5

3.7 Wave Structure

Wave forms in noctilucent clouds (see Fig. 3.9) were first noticed and described in detail by Jesse (1887, 1891) using observations made during the years 1887–1896. He described his observations (Jesse 1890) as follows:

We see a special peculiarity in the structure of the single parts of the clouds against the direction of motion . . . that especially for the very large motions the stripes appear to be structured in such a way that longer, parallel edges lie in the direction of motion, whereas at right angles to them a greater number of shorter crests occur.

Fig. 3.9. Noctilucent cloud system with turbulent variations and wave and billows systems

Störmer (1935) measured a number of cloud photographs taken for triangulation from two or more sites. The clouds seen on June 30, 1934 showed:

The velocity of 80–83 m/s⁻¹ was from east to west. A series of waves with their crests oriented north and south appeared, the distances between successive crests being 6–9 km.

The classification of noctilucent cloud types, principally through the wave structure seen in them, has been described above (Sect. 1.4; see Fig. 3.9).

Grishin (1955) described sub-classes consisting of small crests (Type III-a), crests (III-b) and wavelike arcs (III-c) and in a later paper (Grishin 1961) published a comprehensive study of the meteorological conditions for the appearance of noctilucent clouds. He suggested that anticyclones at ground level can excite oscillations in the upper atmosphere. The oscillations cause adiabatic cooling of sufficient magnitude to lead to the formation of noctilucent clouds. The evolution of the anticyclone is correlated with location and displacements of bands (I-, II- and III-c type waves) of the noctilucent cloud display.

Grishin (1970) and Schröder (1967) published a case study of the noctilucent clouds of July 3–4, 1967. They showed that abrupt changes in velocity or direction are not necessarily real changes; they may result from the effects of perspective or of changes in the slant optical path through an optically thin cloud layer.

Grishin and Kurilova (1973) have discussed observations of noctilucent clouds over the period 1950–1970. Their analysis shows a wide spectrum of wave formations and they suggested that the waves are of different origins. Type III-a are caused by wind shear in the mesosphere. Type III-b may well be generated in the troposphere, and may penetrate upwards into the mesosphere as planetary waves. Type III-c may be caused by mesoscale planetary waves and can control the perceived extent of a noctilucent cloud display.

Witt (1962) has published a comprehensive and detailed study of the morphology of noctilucent clouds. He found the average amplitude of the wave structures to be about 2–3 km, with greater amplitudes in restricted parts of the display. He pointed out:

Wave motion on a smaller scale is represented by systems of parallel billows of different orientations. The appearance of the billows may remind an observer of that of tropospheric clouds or a 'mackerel sky', as sometimes seen in the literature. Analysis shows that, at least in this case, the cloud surface was continuous between the billow crests . . . Thus, it is reasonable to conclude that the increased brightness of the crests is due to geometrical effects rather than to a sublimation and re-evaporation process . . . The characteristic wavelength of the billows is about 4 to 9 km with amplitudes of about 500 to 1000 m.

An analysis of North American data has been published by Fogle (1966) and discussed in detail by Fogle and Haurwitz (1969). They pointed out:

These data show that noctilucent cloud bands are characterized by wavelengths of 10–75 km, amplitudes of 1.5–3 km, lifetimes of several hours, lengths of hundreds of kilometers, and apparent speeds of 10–30 m/sec. The noctilucent cloud billows are characterized by wavelengths of 3–10 km, amplitudes of 0.5–1.0 km, lifetimes of 6–24 min, and lengths of 10–40 km. Billows generally move with the display while the bands generally move in a different direction.

Zhukova and Trubnikov (1966) found that the longer waves may move in a direction opposite to the wind, while the shorter waves move with it. For the billow clouds, Witt (1962) found lengths between 4 and 9 and 30–40 km for the short-period systems.

Fogle and Haurwitz (1969) argued for the appearance of waves at an interface (which is presumably the mesopause) and which show up as noctilucent cloud billows (type III). Another possibility is the presence of internal gravity waves which will exist without the necessity of an interface and which are generated by tropospheric disturbances. Rosenfeld (1986) concluded from his analysis that the waves in noctilucent clouds are part of a single wave process of internal gravity wave propagation.

Hines (1959) realized that internal gravity waves can cause the patterns in noctilucent clouds. Theoretical investigations show that the long-period waves can indicate the presence of internal gravity waves. Hines (1968) described the case for internal gravity waves being the source of the wave forms in noctilucent cloud; he uses the data of Witt (1962) from the night of August 10, 1958. Hines related these measurements to surface frontal systems and interpreted the long-period waves as being caused by internal gravity waves and suggested a possible tropospheric origin in connection with mountain-lee waves and jet streams.

The appearance of the noctilucent cloud billows is similar to that of billow clouds in the troposphere, where a number of fairly equidistant rolls are arranged in parallel lines. This array of cloud rolls has been discussed by

Helmholtz (1889a, b) and is due to wave motion at an interface between two layers (cf. Ertel 1938, 1939). Haurwitz (1931, 1961, 1964) and Haurwitz and Fogle (1969) have discussed in some detail the application of the tropospheric theory to the case of noctilucent clouds which occur in the mesosphere or thermosphere.

In the literature the similarity of noctilucent cloud wave systems to those in cirrus clouds has repeatedly been noted (Wegener 1912; Süring 1943). It has been suggested that the waves seen in noctilucent clouds can be interpreted like those of the cirrus clouds. The waves observed in the high atmosphere thus may represent boundary waves, with a wavelength depending on the wind discontinuity. A wavelength of about 6 km would require a discontinuity of 39–64 m s^{-1}; a 28 km wavelength 208–261 m s^{-1}. It can, however, be assumed that in the mesopause region wind discontinuities of as high as 200 m s^{-1} are rare and it is hardly feasible that the long-period wave systems could be caused by local wind shear while the short-period waves systems could be boundary waves.

A critical summary of wave data in noctilucent cloud displays during the years 1967–1972 has been presented by Ordt and Brodhum (1974). Assuming that internal gravity waves might be involved, the site of suspected sources was predicted by ray-tracing internal gravity waves and subsequently trying to identify the source regions at the 300 mb level. The results lead to an indication that the waves are generated in the tropopause jet stream.

Kuhnke (1976) analyzed the kinematics of the noctilucent cloud display of August 10–11, 1958 and he found wavelengths of 5–8 and 25–30 km. He suggested that the short waves can be interpreted as wind shear waves, while the long waves show the features of internal gravity waves, generated in the tropopause jet stream.

These atmospheric sources must show energy propagation up to the mesopause. The upward propagation has been discussed by Eckart (1960) and calculations show that the short-period waves (periods below 8 min) cannot reach the height of noctilucent clouds. They are reflected or damped at heights above 65 km as a result of the very complex refraction process resulting from the atmospheric temperature profile. A similar finding concerns waves of shorter wavelengths, implying that short-period waves are generated either within noctilucent clouds or at similar heights.

To determine whether a tropospheric jet stream may be considered as the source of a long-period wave structure, the meteorological conditions have to be taken into account. Hines and Reddy (1967) have studied the influence upon gravity wave propagation of winds at heights between 20 and 85 km. By assuming realistic wind and temperature profiles, they determined transmission coefficients for different angles (0°–180°) between the horizontal direction of wave propagation and the direction of wind. An isotropic source was assumed, with the period and phase velocity of the waves being specific parameters of the calculations.

Haurwitz and Reiter (1974) dealt primarily with stationary lee waves generated by orographic obstacles in the troposphere. A reference in the last paragraph to noctilucent clouds is largely an afterthought. They considered whether, on the basis of the numerical estimates made in the main part of the paper for the tropospheric waves, it is possible for internal gravity waves to propagate to the noctilucent cloud heights. It had already been shown by Charney and Drazin (1961) that, especially in summer, the long planetary waves of the low atmosphere would in general not propagate into the high atmosphere.

The effect on the dynamics and chemical composition of the mesosphere and lower thermosphere of gravity wave breaking has been discussed by several authors (Garcia and Solomon 1985; Jakobs et al. 1986). The role of turbulence around the turbopause has been analyzed by Danilov (1984).

Leovy (1966) has studied the possibility of local excitation of internal gravity waves by radiative heating and cooling. He found that both heating by ozone absorption of solar radiation and the heat released during recombination of atomic oxygen may play a role in the growth of the wave but the growth rate would be small. The same problem has been studied by Ivanovsky (1966). Trubnikov and Skuratova (1966) examined the possibility of cellular convection taking place in noctilucent clouds.

In a recent paper Taylor (1986) points out that airglow structures (principally in the hydroxyl emission) often exhibit an appearance similar to noctilucent cloud "bands and billows" and suggests both are manifestations of the same wavelike perturbations.

In summary, it must be admitted that it is still open to question under which conditions the short and long waves in noctilucent clouds are formed; it is important to study the dynamics of the wave motions with regard to temperature fluctuations (see also: Batten 1961; Chapman and Kendall 1965, 1966; Charlson 1965; Chvostikov and Megrelisvilli 1970; Ertel 1938, 1953c; Gadsden 1986c; Grishin 1967b; Keegan 1961; Kohl 1972; Martynkevic 1971, 1973; Scott 1972; Taubenheim 1975).

4 Spectrophotometry

4.1 Introduction

In general, the spectrum of noctilucent clouds shows no features that are not present in the incident sunlight. The solar Fraunhofer lines are present and there are no obvious emission lines. The relative spectral distribution in the scattered light differs from that of sunlight and, in principle, this difference can be measured and used to estimate the size of the scattering particles. In practice, this is neither easy to do nor exact in application. Knowing how to correct for the effect of atmospheric absorption is the nub of the problem of interpreting these observations.

For example, the data (Gadsden 1975) shown in Fig. 3.6 were analyzed by assuming that the amount of ozone to the north of Aberdeen was what is expected at that time of year. It was assumed also that the transmission of the atmosphere, both near the sunset line far to the north of Aberdeen as well as locally, was that given in standard textbooks on astronomy. Then, after allowing for atmospheric absorption on the paths of sunlight up to the particular clouds (assuming that they were at 82-km height) and the transmission of the atmosphere at the oblique directions of observation, the spectral reflectance of the clouds was found to be lower in the blue than would be expected if the cloud scatterers were small in size (Fig. 4.1).

Accepting these data as they stand, the scatterer size is found to be approximately 0.3 μm in radius for a refractive index of 1.33, with a column density of approximately 4×10^7 m^{-2}. Clearly, however, the analysis stands or falls on the precision of allowing for atmospheric absorption and it is difficult to see how this can be done in a rigorous way given the unusual conditions of observation.

The analysis is hindered, first, by the presence of at most two or three stars visible in the twilight arch, and it is not practicable to estimate atmospheric transparency on these occasions from stellar observation. Second, the brightest clouds always appear fairly close to the horizon and atmospheric absorption along the line of sight is large. Third, except in rare cases, there are no simultaneous ozone-sonde observations from a meteorological station some hundreds of kilometres from the noctilucent cloud observatory in the direction of the Sun. An average ozone distribution, appropriate to the season and to the latitude, has to be assumed for the analysis of the spectral data.

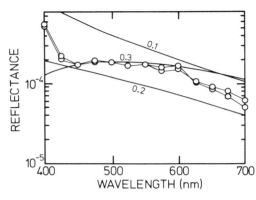

Fig. 4.1. Spectral reflectance plotted (*open circles*) through the visible spectrum for two separate bands in a noctilucent cloud. *Superimposed lines* indicate what would be expected for a monodisperse assembly of scatterers with radii equal to 0.1, 0.2 and 0.3 μm (Gadsden 1975)

It may be that little weight should be given to estimates of particle radius based on ground-level observations of the spectral distribution of the scattered light. As has been mentioned in Chapter 3, rather more emphasis is placed upon estimates of the size of particles from the measurements of the degree of polarization in the scattered light. The degree of polarization will not be changed by passage of the light down through the lower atmosphere after scattering from the clouds. If one knows the state of polarization of the light (usually this is assumed to be unpolarized) illuminating the clouds and one knows the optical properties of the cloud particles, measurement of the polarization of the light from the clouds can give estimates of the particle sizes in the clouds. In such an analysis, many of the problems associated with knowing the atmospheric absorption and refraction on any particular occasion are avoided.

4.2 Spectroscopic Observations

From the time of the earliest observations, spectroscopy has been used as a tool to distinguish noctilucent clouds from auroral structures. The year after Jesse had drawn attention to the unusual nature of the "silver-blue cloudlets", Smyth (1886) gave the following attractive description of an evening sky seen from Edinburgh:

On issuing, then, that night [July 27, 1886], close upon twelve o'clock, from the Observatory computing-room, upon the Calton Hill, I was surprised and even startled, not at seeing a low-down coloured twilight in the north, but at the excessive strength, and glittering brightness of its colours. You might indeed have, at first sight, imagined that some great city, spread abroad over the plains of Fife was in a fierce state of extensive conflagration, so burning red was the first and lowest stratum extending along nearly 20° of the horizon. But that awful kind of redness passed quickly into lemon-yellow clouds in the stratum next above the red; and then came the silver-blue cloudlets just above the lemon-yellow, and even brighter still; but with an innocence of colour and gentleness of beauty, which at once exorcised the horrid idea of malignant flames devouring the works of man; and showed it must be something very different.

. . . At the same time a few stars were faintly visible; while a long streamer, of apparently white cirrus cloud, trailed over half the sky from west to east-north-east, and passed across the Polar region at a considerable altitude, having the silver-blue cloudlets and their gorgeous red basement far below, but within, its wide-enclosing sweep.

(Perhaps Smyth being Astronomer Royal for Scotland helped the editor to accept his letter in spite of it being in a literary style not likely to be acceptable today in a scientific journal.) Smyth used a "a large spectroscope" at his home to observe the different parts of the sky. The red glow close to the horizon showed, he thought, an emission line in the green which, by micrometer measurement and reference to a hydrogen lamp, he found was "in the very position of" the auroral green line.

. . . in fact, aurora was at that moment assisting, though to a very small extent, in that low streak of merely, but yet so intensely coloured, solar and Scottish, midsummer-midnight, northern, twilight.

With a hand spectroscope, he could see that the band of "cirrus cloud" was simply an auroral band or arc. He saw

the aurora's chief line thin, sharp and positively brilliant along its whole extent.

Later in the night, about 1 AM, the auroral band brightened, began to twist and many auroral rays developed. In fact, it was a "very fair" auroral display.

4.3 Spectrophotometry from Ground Level

Grishin (1956) published the first photographic spectra of the clouds. His spectrograph was a prismatic one and the spectra are rather crowded in the red but the minimum in the yellow (the Chappuis absorption) appears. Deirmendjian and Vestine (1959) discussed the analysis and significance of Grishin's data in some detail and concluded that there is primary scattering of direct sunlight by dielectric spheres with a maximum radius of 0.4 μm together with some night sky emission lines and bands.

Fogle and Rees (1972) were able to obtain spectra on two nights; on one, the cloud intensity was too low for a good analysis. The other night, July 22, 1969, gave usable spectra only in the 300–500 nm region.

One of their spectra is shown, redrawn, in Fig. 4.2. Fraunhofer lines show up clearly in the violet part of this spectrum. Calibration of the photoelectric

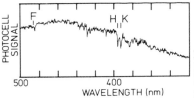

Fig. 4.2. Spectrum of a noctilucent cloud observed by Fogle and Rees (1971) on July 22, 1969. The H and K Fraunhofer lines at 396.8 and 393.4 nm are clearly present, with perhaps the F line at 486.1 nm

response showed that the radiance of the noctilucent cloud at 420 nm was close to 4 kR/A. (This figure is not corrected for slant viewing; making the correction reduces the radiance to 0.9 kR/A for vertical viewing.)

The measurements have been compared with the scattering predicted from a cloud consisting of ice particles with a radius of 0.13 μm and gave good agreement with a column density of just under 10^9 m^{-2}. Fogle and Rees (1972) corrected the data for both the twilight foreground light and for atmospheric absorption before making this estimate.

Harrison (1973) measured the spectral radiance in the near-infrared of a single cloud seen north of Calgary (Canada) on June 14, 1972. The display was the only one to appear in a series of observations throughout June and July when data for airglow and auroral spectra were being obtained. The spectral regions covered were 785 to 900 nm and 2.8 to 4.2 μm. Spectra were also obtained regularly of the (4–1) band of hydroxyl, which lies in the 1.02 to 1.064 μm region.

Harrison found that the spectrum of the scattered radiation in the red and near-infrared from the noctilucent cloud showed no strong absorption features. At longer wavelengths, there was little evidence of scattered sunlight; the thermal emission from the atmosphere was predominant. Harrison pointed out that this is to be expected unless the radius of the scatterers was greater than 1 μm.

Veselov et al. (1976) reported on a series of measurements made in the summer of 1974 in the spectral range 400 to 1700 nm. Because of unfavourable weather during the summer, measurements were possible only on the rather faint display of July 9, 1974. The results show a monotonically decreasing spectral radiance in the range covered; the scattering angle for the observation was approximately 20°.

4.4 Rocket-Borne Photometers

It appears that Witt (1969) was the first to project a photoelectric photometer through a noctilucent cloud. He found a distinct layer, with a sharp lower boundary, on both ascent and descent of the rocket. There were problems with the illumination of the photometers by the Sun but data were successfully obtained. The effective wavelength of the radiation was 366 nm on one channel and 534 nm on another. The scattering angle was 86°. The layer was entered at 82.4 km altitude, and the rocket seems to have broken clear of the layer by about 84.4 km. The ratio of the radiances at the two wavelengths (534:366 nm) was at least 2.5. There was a high degree of linear polarization, and if the cloud particles were scattering as monodisperse dielectric spheres of refractive index 1.33, an upper limit of 0.2 μm in radius was indicated. The column abundance of such particles was found to exceed 10^8 m^{-2}.

Witt was involved in another rocket sounding from Kiruna (69°N) the following year (Witt et al. 1976). For this flight, the wavelengths chosen for the photometers were 256 and 536 nm. Both instruments sensed a scattering layer at altitudes between 85.5 and 89 km on both ascent and descent. The 256-nm signal was affected to a considerable extent by fluorescence from upper atmosphere nitric oxide. The strength of the scattered light at 536 nm from the layer amounted to roughly one-half that from molecular scattering at 85.5 km. The data do not refer to noctilucent clouds but rather to a scattering layer somewhat higher than where the clouds are usually seen. Witt et al. deduced an upper limit to the particle radius of 0.05 μm. These may well be the particles available for nucleation of noctilucent clouds when the temperature or humidity is suitable. The scattering would have been caused by a column abundance of approximately 10^9 m^{-2}; for a layer thickness of 3 km, the mean particle density becomes 3×10^5 m^{-3}.

Witt et al. (1971) presented a report on two flights from Kiruna in 1971 using more sophisticated photometers. These involved measurements at seven wavelengths (214, 309, 366, 453, 536, 589 and 762 nm) and included a number of other instruments (airglow infrared photometers, nitric oxide ionization chamber, solar irradiance photometer, ion collector and Faraday rotation). The preliminary results from the photometers showed noctilucent clouds on both flights. On the July 31 flight, the layer was between 82 and 83 km; the following night, the cloud layer was higher, i.e. 85–87 km.

Rössler (1972) has made measurements of the scattered radiance in a cloud from a rocket launched from Kiruna at 0040 UT on Aug. 10, 1970. There was clear evidence of a scattering layer at 82.5 km with integrated radiances (at 470 and 600 nm) of 0.1 W m^{-2} sr^{-1}. A flight on Aug. 14, 1969, "in the case of no noctilucent clouds", had shown a scattering layer at 83 km altitude with the same radiance at a wavelength of 460 nm. Rössler fitted his data to a model containing two classes of scatterers, Aitken nuclei of radii between 0.03 and 0.05 μm and particles ("Junge particles") with larger radii in the range of 0.1 to 1 μm. The derived number of particles in all cases is close to 500 m^{-3} for the larger particles and 10^6 m^{-3} for the smaller, Aitken particles. De Bary and Rössler (1974) returned to the analysis of the data from the Aug. 10, 1970 flight (the one with a cloud present) and obtained the results listed in Table 4.1 for scattering at three heights, viz. just above, in and just below the cloud. There are differences, as one would expect, between the relative numbers of the two groups of particles at the different heights.

As part of a campaign to launch particle collectors through noctilucent clouds, Tozer and Beeson (1974) obtained photometric data from four flights

Table 4.1. Results from a rocket flight on August 10, 1970

Height (km)	Scattering angle (deg)	Particle numbers (m^{-3})			
		600 nm		430 nm	
		Aitken	Larger	Aitken	Larger
87	85	2×10^4	40	5×10^3	50
82	77	5×10^5	250	3×10^5	500
74	77	4×10^5	30	4×10^5	150

over Kiruna. The instrument on each of the flights had two channels sharing a common entrance pupil (through the use of a split fibre optic bundle). Filters gave effective wavelengths of 540 and 410 nm.

The published data include upward-looking measurements of radiances at 250-m intervals at heights between 60 and 101 km. The better data were obtained from the last two flights which were separated by 31 min around midnight (local time) on the night of August 1, 1973. On both of these flights, the rocket passed through a bright noctilucent cloud which was located between 83.3 and 84.4 km. On all four flights (i.e. including the first two in the series, on July 31 and August 1, 1971), a weakly scattering layer was seen low at approximately 68 km altitude. [This confirms the observations of Fiocco and Grams (1969), who detected layering in this height region with a ground-based laser.] There was no penetration of a noctilucent cloud on the August 1, 1971 flight. Tozer and Beeson summarized their data by suggesting that the cloud particles are principally less than 0.13 μm in radius, with a number density in the cloud of approximately 10^6 m^{-3} (and thus a column density of 1.1×10^9 m^{-2}). Their measured radiances are listed in Table 4.2 and are in accord with the (corrected) radiance measured by Fogle and Rees, quoted earlier in Section 4.3 (that is, 0.9 kR/A at 420 nm).

Gadsden (1978) has scaled the data from the 1973 flights plotted in the Tozer and Beeson paper. By numerical differentiation, volume-scattering coefficients were deduced for altitudes from well below to well above the cloud layer. The 410-nm data in particular gave very convincing scattering data and comparison with expectations based on molecular scattering in the atmosphere below the cloud layer was good. The rocket data can be used to obtain the ratio of the radiances at the two wavelengths, and the derived ratio of the volume-scattering coefficients for the region below the cloud was found to be 0.30, in good agreement with the value 0.33 expected from molecular scattering. The scattering in the cloud layer gave a ratio, using the same data, of 0.33, exactly the same as that for molecular scattering. This confirms that there is no significant proportion of large particles (> 0.2 μm) in the cloud.

Table 4.2. Results from rocket flights over Kiruna

Date	Wavelength (nm)	Radiance (kR/A)
1971 Jul 31 Ascent	540	0.072
	410	0.075
Descent	540	0.079
1973 Aug 1 (A) Ascent	540	0.20
	410	0.45
(B) Ascent	540	0.34
	410	0.80
Descent	540	0.33
	410	0.68

4.5 Spectrophotometry from Satellites

Visual observations on board the USSR and the USA space laboratories have demonstrated that noctilucent clouds can be seen from above. On one orbit of Voshkod-2 (Lazarev and Leonov 1973), a sharply defined, blue-grey layer was seen at 1.5° distance above the nighttime horizon. The apparent height of the layer was deduced to be 80 km and there was perceptible absorption of starlight passing through the layer.

Packer and Packer (1977) gave a brief report of some visual observations by Weitz on Skylab-2. Noctilucent clouds were seen on four occasions during the 28-day flight by Weitz; they were seen always at dawn and in the direction of the rising Sun. They were never seen after the Sun had risen. Packer and Packer reported:

The clouds were bright, as conjectured, and formed a thin bright line just above the earth horizon when first detected. As the spacecraft approached, the thin line appeared to become broken, and finally two to four patchy, thin, stratified clouds were visible. Their lateral angular subtense was of the order of 5°, and as the spacecraft drew nearer, the clouds appeared to rise above the earth horizon, finally vanishing into the airglow. The sightings were made . . . during the last week in May and the first week in June 1973 near 50° north latitude and 10–40° east longitude.

This particular astronaut was impressed by the spectacular views of the horizon and especially by those at twilight. He spent as much time as his duties permitted in watching the view.

The cosmonauts on the Salyut series of spacecraft have paid a considerable amount of attention to observation of noctilucent clouds. Willmann et al. (1977) reported 27 occasions when the cosmonauts reported noctilucent clouds. They were the same clouds reported from ground level in widespread sightings from northern Europe and the northern USSR at the times of passage of Salyut-4. Views both edge-on (Fig. 4.3) and showing cellular (horizontal) structure were obtained from the spacecraft.

In 1972, Donahue, Guenther and Blamont (1972) announced the discovery of a thin scattering layer over the poles of the Earth which appeared to have a mean altitude of 84.3 km (Donahue and Guenther 1973) and to be a summer-time phenomenon. The similarity to noctilucent clouds was obvious and the identification was made immediately by the discoverers.

Donahue and Guenther reported some interesting statistics of the height of the layer. The average is 84.3 km and there is some indication that this progressively decreases as the summer season progresses (with a drop of 1.5 km in the 3 weeks before the solstice). There is a systematic dawn:dusk variation at 70–75° latitude. The average of the measured heights was 85.3 ± 0.8 km in the afternoon (local time 1410–1630 and 82.7 ± 0.6 km in the morning (0512–0538).

The layer was detected and measured with a photometer on board the orbiting geophysical observatory OGO-6. The photometer was designed to measure the airglow layers of atomic sodium and atomic oxygen (the

Fig. 4.3. A noctilucent cloud photographed from space. The noctilucent cloud is seen as a thin, bright line at the top of the illuminated atmosphere. This photograph was taken in September 1985 from the Salyut-7 spacecraft. [Reproduced here through the good offices of cosmonaut G.M. Grechko, Institute of Atmospheric Physics (USSR Academy of Sciences), Moscow.]

589.0 ± 589.6 nm and the 557.7 nm emissions). It was a horizon scanning instrument, with a field of view corresponding to about 5 km in height at an altitude of 100 km. The scattering layer was seen on all satellite passes above 80° latitude from June 5, 1969, the day of launch. The signal appeared to come from a thin layer which was clearly separated from the airglow layers (higher in the atmosphere) and was much more intense than was expected of an airglow layer. The signal stood out clearly above the scattering from the atmosphere just above and just below the layer; from the examples (redrawn in Fig. 4.4) given in the paper, it looks as though the scattered intensity from the layer at 84 km is equal to that from the atmosphere some 20 km lower.

Analysis of the radiance suggested that the layer had a vertical optical depth of approximately 5×10^{-5}; if the scatterers were ice particles of spherical radius 0.13 μm, the authors calculated that there must have been approximately 6×10^{10} m^{-2}. This is a 100 times or more than is seen in noctilucent clouds at lower latitudes and the suggestion of Donahue et al. is that clouds seen at lower latitudes from ground level are a weak, sporadic manifestation of the persistent polar layer. They observed the recurrence of the layer over the north pole in the summer of 1970 and its appearance over the south pole in late 1969.

Hummel (1977) reported that the concentrations of scatterers estimated by Donahue and colleagues need to be increased five to seven times because an incorrect scattering cross-section was used in the calculations.

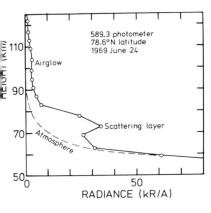

Fig. 4.4. Sample of data showing the scattering layer detected by the sodium airglow photometer (λ = 589.3 nm) on OGO6, redrawn from Donahue, Guenther and Blamont (1972). The *broken line*, labelled "*atmosphere*", is an extrapolation from the radiances measured for heights between 40 and 60 km. The height scale shows the height of the tangent above the Earth's surface

However, Thomas (1984) reported that Donahue found an error in his estimates of the scattering cross-section (which were taken by simple scaling of radiance from the results of Fogle and Rees 1972) amounting to a factor of ten. The column density is 6×10^9 m^{-2}, a better agreement (although still high) with ground-based estimates. The status of Hummel's factor of seven (which should presumably now be increased to 70) is not clear.

Hummel (1975) and Hummel and Olivero (1976) have used the data from OGO-6 to confirm that the ice crystals had radii no larger than the assumed radius of 0.13 μm. Their estimate of radius was made by taking the ratio of radiances of the scattering layer at the two (airglow) wavelengths available from the photometer data. The data are shown in Fig. 4.5.

Gadsden (1978) has attacked this conclusion. There is too much spread in the measured ratio to allow any worthwhile conclusions to be drawn from these data. The great scatter in the plotted points is due to the characteristics of the photometer; each channel, the oxygen and the sodium photometers, viewed the layer sequentially, not concurrently. The interval between consecutive observations at each wavelength is 74 s; taking a ratio of the radiance at one wavelength to the mean of the radiances at the other wavelength measured just before and just after involves comparing three distinct regions of the scattering layer separated through satellite motion over a line approximately 600 km in length along the track. Thus, the layer is unlikely to show uniformity of radiance.

Thomas (1984) reported data from the Solar Mesosphere Explorer (SME) satellite (Fig. 4.6) taken with an ultraviolet spectrophotometer. At wavelengths of 265 and 296 nm (with a band pass of 1.5 nm at each wavelength), measurements of radiance are recorded every 0.8° along the orbit. The instrument is a horizon-scanning device rotating once every 12 s. In each rotation, 32 data samples are taken at each wavelength and the samples are separated by 3.5 km in height at the tangent point. A sample of the observations is shown in Fig. 4.7, which has been redrawn from one of the diagrams presented by Thomas at the Tallinn Workshop on NLC (Thomas and McKay 1986). Note the greater effect

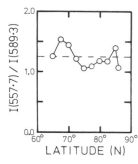
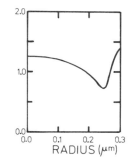

Fig. 4.5. *Left* OGO6 data used by Hummel and Olivero (1976); the ratio of 589.3 nm radiance to that at 557.7 nm is plotted for a number of latitudes of the point of observation. The *broken line* marks the ratio expected for very small scatterers. *Right* The ratio of radiances calculated from Mie scattering theory and plotted against the radius of the scatterer

Fig. 4.6. Montage of the Solar Mesosphere Explorer (SME) satellite. (Reproduced with permission of G.E. Thomas, Laboratory for Atmospheric and Space Physics, Colorado University, USA.)

of ozone absorption at the shorter (265 nm) wavelength. Note, too, how high the scattering layer appears to be on this occasion.

Thomas and McKay (1985) estimated the contribution of Rayleigh scattering from the atmosphere to be as shown by the lines in Fig. 4.7. The scattering layer appears as an excess of radiance above the Rayleigh-scattered component. The layer appears to extend down over a height interval larger than the real layer thickness because these are *limb-scanning* measurements and the observed

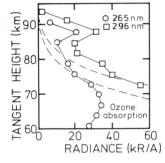

Fig. 4.7. Sample of data showing the scattering layer detected by the UV spectrometer on SME. Latitude 77°N, June 17, 1984 (Thomas and McKay 1985 1986)

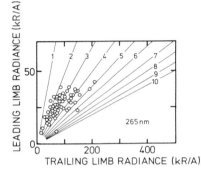

Fig. 4.8. Observed ratio of forward- and back-scattered radiance at λ = 265 nm. Only those data with coincidence of field of view to within ±17 km are shown. Lines of constant ratio are drawn behind the data points (Thomas and McKay 1985 1986)

radiance is set by the length of the line of sight in the layer above the particular tangent point.

The SME spacecraft was spun in a "cartwheel" fashion and Thomas was able to observe the horizon in both the forward ("leading limb") and backward view ("trailing limb"). He thus obtained observations at two scattering angles (132° ± 6° and 48° ± 6°) for each wavelength.

The 265-nm data from orbits in a short period in the (northern hemisphere) summer of 1983 are summarized in Fig. 4.8. It can be seen that there is a considerable degree of forward scattering present, which becomes more obvious as the brightness of the scattering layer increases.

Calculations of the ratio of forward scattering to back scattering (I_{48}/I_{132}) are straightforward if one uses Mie scattering theory. The scatterers are assumed to be spherical with a refractive index of $1.336 - 5 \times 10^{-8}$ i (appropriate to ice at 130 K and λ = 265 nm). The result is shown in Fig. 4.9.

By the time the radius of the scattering sphere has grown to approximately λ/π in size, there is appreciable forward scattering present. As the radius increases still more, a series of deep minima in the back-scattering half of the scattering diagram develops; these minima move, in angle, progressively forward as the radius continues to increase and at the same time the minima become less deep. This is shown clearly in Fig. 4.9 by the very obvious peaks (two of them off-scale) in the ratios plotted for monodisperse assemblies of scatterers.

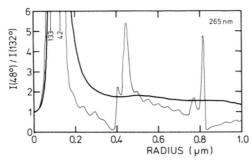

Fig. 4.9. Calculated ratio of the forward- and the back-scattering 265–nm radiances for radii up to 1 μm. The *thin line* refers to a monodisperse assembly of scatterers; the *thick line* denotes an assembly with a size distribution which is constant up to an upper limit of radius. The first two peaks in the ratio occur at r = 0.085 μm (*133*) and 0.130 μm (*42*). For the illustrative size distribution, these peaks become 23 and 27 respectively

Clearly, real noctilucent clouds do not consist of ice crystals that are spherical and uniform in size; there is a spread in sizes and the rapid changes in ratio that are shown in Fig. 4.9 will be smoothed out in practice. As the characteristic size of the cloud crystals increases, the observed ratio will go through much less extreme variations. For illustration, the heavy line in Fig. 4.9 shows what will happen if the size distribution shows numbers of scatterers that are constant for every interval in radius up to an upper limit. Because the scattering efficiency of a scatterer at first increases very quickly with an increase in radius while settling down at larger radii towards an r^2 increase, the first peaks in radius remain in the heavy-line plot while the later ones become smoothed out.

A second, independent quantity for comparison with scattering theory is the ratio of the radiances at 265 and 290 nm. The ratio in sunlight is quoted by Thomas and McKay (1985) as 0.35; if this number is multiplied by the ratio of scattering efficiencies at the two wavelengths (the curves shown in Fig. 4.10), the ratio of radiances is obtained. In calculating the scattering efficiencies for Fig. 4.10, the refractive index at 290 nm was taken equal to $1.328 - 6 \times 10^{-9}i$.

Clearly, the ratio of radiances will be less than 1.0 unless the scatterers have characteristic radii of either 0.1–0.2 μm, or 0.5μm. The SME data show a slight fall in ratio, from 1.4 (132°) and 1.5 (48°) to 1.2 and 1.35, for observations with I_{48}/I_{132} ratios changing from between 1 and 2 to approximately 5. The scatter in the observations is so great, however, that it is not possible to be dogmatic about the characteristic size of cloud particles and it may well be that the ratios plotted in Thomas and McKay (1985: Fig. 11) need adjustment.

Olivero and Thomas (1986) have studied the SME data obtained from observations made between December 1981 and September 1984. Polar mesospheric clouds were seen in the summers of each hemisphere. Olivero and Thomas reported that the brightest layers are found at latitudes where they are detected most frequently (70°–75°) and that the "seasons" for the clouds last

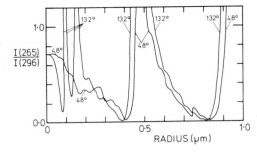

Fig. 4.10. Calculated ratio of scattering efficiencies, I_{265}/I_{296}, for the two scattering angles (48° and 132°) relevant to the SME data

approximately 70 days, centred close to 20 days after the solstice. During the 3 years considered, the northern hemisphere clouds were statistically brighter than the southern hemisphere clouds.

To turn now to data from other spacecraft, there were four near-infrared photometers on Salyut-4 available for horizon scanning (Avaste et al. 1977; Avaste and Keevallik 1982). The effective wavelengths were 1.35, 1.9, 2.2 and 2.7 μm; the photometer at the shortest wavelength (1.35 μm) showed a plateau in the plot of radiance against distance above the horizon, while there was no indication of noctilucent clouds at 1.9 μm. On another occasion, the 1.9-μm channel showed a clear peak in radiance corresponding to a layer at about 80 km. The relative and absolute radiances at 1.35, 1.9 and 2.2 μm showed, from observations for a scattering angle of 80°, that the vertical optical thickness of the clouds was in the range 5×10^{-5} to 10^{-4}.

Belyaev et al. (1979a) reported on measurements of the spectral radiance of noctilucent clouds in the visible region. The data came from a small spectrometer on board Salyut-4 with a rapid scan in wavelength. Measurements were obtained as the spacecraft flew on a track between 40° and 50°N latitude past noctilucent clouds at around 60° latitude. The authors show a plot of relative radiance from 400 to 800 nm for three scattering angles, 22° ±7°, 47° ±8° and 51° ±5°. In the red, at about 740 nm, the radiances are in the ratios 1.0:0.71:0.46; at 400 nm, the ratios are quite different, 1.0:0.47:0.26. It can be noted that the ratios are not in the proportions of $(1 + \cos^2\theta)$ but there could well have been problems in sighting the entrance slit of the spectrometer during a spectral scan; the ratio of radiances is therefore not necessarily evidence of a forward-scattered component in the scattered light.

The same data are discussed in a later paper by Belyaev et al. (1979b). They noted that the data are uncorrected for absorption by the lower atmosphere of the sunlight incident upon the cloud and that the field of view of the instrument was larger than the projected area of the clouds. Belyayev et al. (1981) returned to these same spectral data and gave absolute calibrations for them. They concluded that there is a satisfactory fit with cloud particles having an effective radius of 0.7 μm, with a cloud density of 6×10^6 m^{-3}. Alternatively, a distribution of particle sizes is considered which is described by the function Ar^n, over a range of sizes from $r_{min} = 0.025$ μm to $r_{max} = 1.9$ μm, with n = −3.5.

4.6 Conclusions About Cloud Particle Sizes

Avaste et al. (1981) described the results of microphotometry of some of the Salyut-6 photographs of noctilucent clouds seen at the horizon, in the direction of the Sun. The data show quite a marked degree of forward scattering in the clouds; there is a clear brightening of the scattered light when the azimuth relative to the solar azimuth is less than 5°. It will be recalled that Weitz's observations on Skylab were also indicative of a marked degree of forward scattering. It is difficult to reconcile this with a model of particle sizes which contains no particles of radius greater than 0.13 μm.

In a review paper, Avaste et al. (1980) summarized much of the USSR data obtained from the Salyut missions and listed the following conclusions drawn from the visual observations and photography from space:

1. Photography and visual observations from space enable one to determine the NLC on a global scale. In particular, the investigations performed on board 'Salyut-4' revealed that in summer in the Northern Hemisphere NLC often completely covered the latitudinal belt north of 45°. Similar observations aboard 'Salyut-6' ascertained that in summer in the Southern Hemisphere there also exist extensive NLC fields, but they are shifted more southward: the NLC belt is south of 53–55°S.
2. Photometric investigations as well as visual observations revealed that in both hemispheres the mesopause often has a complex structure (sometimes there exist two- and three-layered NLC fields).
3. These observations allow one to make sound estimates of the spatial-temporal characteristics of NLC fields as well as the morphological features of their evolution.
4. The photometric investigations from space also confirm that in the NLC layer there exist particles whose radius exceeds 10^{-5} cm. The NLC in the Southern Hemisphere probably consist of smaller particles than those in the Northern Hemisphere.

It is instructive to bring together the various conclusions reached by different observers (Table 4.3). There seem to be two separate schools of thought. One, the "Western School", feels that the data demonstrate that there are only small cloud particles present. The upper limit on the scatterer radius is at most 0.13 μm and is more likely to be half this size. The other school, the "Eastern", relies on observations in the visual and the infrared to assert that there are particles present whose radii are a large fraction of a micrometer (0.3 to 0.7 or 0.9 μm). It will be suggested in Chapter 5 that the data on polarization of the light scattered from noctilucent clouds also offer no definite conclusions. In situ sampling (Chap. 6) seems to provide support for the Eastern, or large particle, school but even these data are now regarded as not inconclusive.

Table 4.3. Estimates by spectrophotometry of cloud particle sizes[a]

Author(s)		$\lambda(\mu m)$	$r(\mu m)$	$N(m^{-2})$	Notes
Grishin (1956)[b]	G	0.55	0.4		
Witt (1969)	R	0.45	< 0.2	> 10^8	
Fogle and Rees (1972)	G	0.4	0.13	10^9	
Rössler (1972)	R	0.53	0.04	10^{9e}	
			0.1–1.0	5×10^{5e}	
Harrison (1973)	G	3.5	< 1.0		
Donahue and Guenther (1973)	S	0.57		6×10^{9c}	
Tozer and Beeson (1974)	R	0.47	< 0.13	10^9	
Gadsden (1975)	G	0.55	0.3	4×10^7	
Witt et al. (1976)	R	0.40	< 0.05	10^9	
Hummel (1977)	S	0.57	< 0.13	4×10^{11d}	
Belyaev et al. (1981)	S	0.6	0.7	6×10^{9e}	
Thomas and McKay (1985)	S	0.28	< 0.06	$10^{10} - 10^{14e}$	

[a] G = ground-based measurements; R = rocket-borne measurements; S = satellite-borne measurements.
[b] Discussed by Deirmendjian and Vestine (1959).
[c] Estimate decreased by a factor of 10 according to the discussion of Thomas (1984).
[d] The same data as that used by Donahue and Guenther (1973).
[e] Assuming cloud layer is 1 km thick.

5 Polarimetry

5.1 Introduction

In principle, the problem of predicting the state of polarization of light scattered from noctilucent clouds is solvable. Given the state of polarization of the incident sunlight, the light that would be scattered from any model containing spheres, cylinders or spheroids can be calculated once the composition and the refractive index of the material (or materials) in the particles is determined. However, the sunlight that is incident on the clouds has passed obliquely through a considerable amount of the atmosphere. By the very nature of observing the clouds in twilight, some time after local sunset, the incident sunlight has entered the atmosphere, plunged fairly deeply over the terminator and then begun to emerge again before striking the clouds. There will be significant absorption along this path. In addition, there is at least the possibility of a significant contribution to irradiance at the cloud level from forward-scattered light in the troposphere and stratosphere. This problem has not yet been solved. It is less of a problem for observations of noctilucent clouds from rockets (where the Sun may be well above the local horizon at cloud level).

The degree of polarization will not be changed by passage of the light down through the lower atmosphere after scattering from the clouds. If one knows the state of polarization of the light illuminating the clouds (usually this is assumed to be unpolarized light) and if one knows the optical properties of the cloud particles, measurement of the polarization of the light from the clouds can give estimates of the particle sizes in the clouds. In such an analysis, many of the problems associated with knowing the atmospheric absorption and refraction on any particular occasion are avoided.

The state of polarization of light is described through the use of Stokes parameters. The transformation of these parameters by scattering can be treated with the help of the Mueller calculus (Shurcliff 1962). In using this calculus, one writes the Stokes parameters as the Stokes vector $\{I,M,C,S\}$ having four components. The effect on the vector of any linear transformation (e.g. that caused by scattering, or by passage of the light through a polarizing element) is expressed by 16 coefficients in a 4×4 matrix, F. If the incident light has Stokes parameters I, M, C, S then the Stokes vector that results from the operation of the linear transformation is simply the product of the matrix F with the Stokes vector of the incident light. That is

$$\{I,M,C,S\} = F \cdot \{I_0,M_0,C_0,S_0\} \ . \tag{5.1}$$

The physical meaning of the individual Stokes parameters can be grasped by referring to Fig. 5.1, a representation of the polarization ellipse. Using the quantities marked on the diagram, we have the following relationships:
Ellipticity $= \tan\beta$; orientation of the major axis $= X$.
The Stokes parameters are

$$I = a^2 ; \tag{5.2a}$$
$$M = I \cos 2\beta \cos 2X; \tag{5.2b}$$
$$C = I \cos 2\beta \sin 2X; \tag{5.2c}$$
$$S = I \sin 2\beta. \tag{5.2d}$$

If $M = +I(-I)$, the light is linearly polarized in (perpendicular to) the scattering plane.

If $C = +I(-I)$, the light is linearly polarized at $45°$ ($135°$) to the scattering plane.

If $S = +I$ $(-I)$, the light has right-handed (left-handed) circular polarization.

The angle X is given by $\tan^{-1}(C/M)$. Clearly,

$$I^2 = M^2 + C^2 + S^2 \tag{5.3}$$

for a fully polarized light beam. The degree of polarization of a partially polarized light beam is given by the ratio

$$\sqrt{M^2 + C^2 + S^2}/I.$$

The principal attraction of handling the analysis of polarized light through the Stokes vector approach is that the Stokes parameters obey the algebraic laws of addition. That is to say if two incoherent beams of light, described individually by the two Stokes vectors,

$$\{I_1, M_1, C_1, S_1\} , \{I_2, M_2, C_2, S_2\} , \tag{5.4}$$

are combined into one beam of light, the Stokes vector of that combined beam will be

$$\{I_1 + I_2, M_1 + M_2, C_1 + C_2, S_1 + S_2\} . \tag{5.5}$$

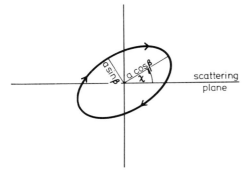

scattering
plane

Fig. 5.1. Representation of the state of polarization of a light beam showing elliptical polarization. The direction of propagation is *into* the page

This property is of special application in observing noctilucent clouds. The clouds are seen in twilight and there is, therefore, an appreciable foreground of light scattered in the lower atmosphere. Measurements are made of part of a noctilucent cloud together with measurement of an adjacent part of the twilight sky, which has no noctilucent cloud behind it. The two sets of measurements are separately reduced to give Stokes parameters. The difference between the two sets of parameters gives the Stokes vector of the noctilucent cloud, freed of the foregound light.

Then follows the analytical stage: given the scattered light, $\{I,M,C,S\}$, from observation and assuming that the light which illuminates the cloud, $\{I_0,M_0,C_0,S_0\}$, is known, can one deduce the scattering matrix, F, involved in (5.1)?

Van de Hulst (1957) and Kerker (1969) discussed the form and coefficients of the matrix F in some detail, with special reference to cases of matrices for scattering media. If the particles in cloud scatter waves, which have unrelated phases (that is, if the scattering in the cloud is incoherent), each coefficient in the matrix F for the cloud is the sum of the corresponding coefficients for each particle. Also, because the scattering is assumed to be incoherent, the summing for the cloud of particles is done for intensities, not amplitudes, of the scattered waves. The principle of optical equivalence enunciated by Stokes in 1852 (see Clarke and Grainger 1971) makes clear that it is impossible to distinguish between different incoherent sums of simple waves that together form beams of light with the same Stokes parameters.

5.2 Polarization by Scattering

The amount of light scattered from a region varies with the size of the scattering angle, θ, and (in the case of non-spherical or anisotropic scatterers) with the azimuth, ϕ, of the direction of the scattered light. These angles and the orientation of the scatterer are measured relative to the scattering plane which is defined as the plane containing the direction of the incident light and the point, P, at which the scattered light is detected. The scattering angle, θ, is zero in the direction of travel of the incident light (forward scattering) and 180° (back scattering) if the scattered light travels antiparallel to the incident light.

The radiance of the scattered light is related to the irradiance of the incident light through a scaling factor $(1/k^2r^2)$ in which k is the wave number $2\pi/\lambda$, and r is the distance between the scatterer and P. Only rarely is the magnitude of this scaling factor needed or used in the analysis of the polarization of light scattered from a noctilucent cloud and it will not be discussed further.

There are certain restrictions that are imposed on the matrix F by the symmetry existing in some physical situations. First, if all the scattering particles are spheres, or if they have a plane of symmetry but their axes are oriented randomly, 8 of the 16 coefficients are zero and two pairs of those that remain share the same numerical magnitude, so the matrix contains only six separate coefficients:

$$F = \begin{matrix} S_{11} & S_{12} & 0 & 0 \\ S_{12} & S_{22} & 0 & 0 \\ 0 & 0 & S_{33} & S_{34} \\ 0 & 0 & -S_{34} & S_{44} \end{matrix} \quad . \tag{5.6}$$

In this case, there are limits to the type of polarization that the scattered light can have. If the incident light is unpolarized, $\{I,0,0,0\}$, the scattered light has Stokes parameters $\{S_{11}I, S_{12}I, 0, 0\}$. It has plane polarization, therefore, either in the plane of scattering or at right angles to it depending on whether S_{12} is positive or negative in value. If the incident light is linearly polarized, the Stokes parameters of the scattered light are $\{(S_{11}I + S_{12}M), (S_{12}I + S_{22}M), S_{33}C, -S_{34}C\}$. There can be circular polarization if the incident light has any degree of linear polarization in a plane oblique to the scattering plane. Naturally, if the incident light is elliptically polarized, that is, if there are both kinds of linear polarization plus circular polarization present in the incident light, the scattered light will, in general, be elliptically polarized too with neither the original orientation of axes nor the original ellipticity being necessarily preserved.

In the general case where the cloud contains odd particles which are not randomly oriented, the matrix F is unrestricted. It should be noted that the coefficients in F are related to both the shape and to the composition of the cloud particles and there is no way of distinguishing particles with anisotropic composition from elongated particles of isotropic material.

There are two simplifying approximations that can be made and often are made. If the particles (of arbitrary shape) are very small in comparison with the wavelength of the incident light, the Rayleigh approximation holds. In this case,

$$F = \begin{matrix} 0.5(\cos^2\theta + 1) & 0.5(\cos^2\theta - 1) & 0 & 0 \\ 0.5(\cos^2\theta - 1) & 0.5(\cos^2\theta + 1) & 0 & 0 \\ 0 & 0 & \cos\theta & 0 \\ 0 & 0 & 0 & \cos\theta \end{matrix} \tag{5.7}$$

with all the consequent simplifications in taking the product with $\{I_o, M_o, C_o, S_o\}$ to find the Stokes parameters of the scattered light.

The second approximation is that the scattering particle is low in contrast, that is, the phase of the incident wave is negligibly changed in passing the particle. Lord Rayleigh (1881) gave this approximation as an extension of the small-particle approximation which is valid for particles of any size and any shape provided their refractive index is close to that of the surrounding medium. To distinguish the two approximations discussed by Rayleigh, the low-contrast case is usually called Rayleigh-Gans scattering, but Kerker (1969) has suggested this should be more appropriately called Rayleigh-Debye scattering. The Rayleigh-Gans approximation is not recommended for the case of noctilucent clouds, where the refractive index of the cloud particles is not close to unity. It is, however, useful in permitting simpler relations to be used to indicate the behaviour of full, rigorous solutions to the scattering problem.

Wait (1955) has obtained the full solution for the case of scattering from an infinitely long cylinder for which, as Rayleigh pointed out in 1881, one may also make use of the equations for a scatterer of finite length. Banderman and Kemp (1973) obtained expressions for the degree of circular polarization produced when unpolarized incident light is scattered from a finite right circular cylinder. They made the approximation that the relative refractive index of the cylinder is close to unity so there is a limit to the usefulness of their treatment being applied to the case of noctilucent cloud particles. Nevertheless, the behaviour exhibited by the low-contrast cylinders can be useful as a guide and has been so used by Gadsden (1977). A full solution of Maxwell's equations for the case of scattering by an ellipsoid has been obtained by Asano and Yamamoto (1975) and this set of equations should be applied when testing a cloud model against observations of the scattered light. The solution, however, involves radial spheroidal wave functions of the first and second kinds and there is not, at present, a computer programme available to do these calculations.

Barber and Yeh (1975) have approached the problem from a different direction and obtained general solutions using the "Extended Boundary Condition Method"; in their work, they give examples of scattering from several different shapes of scatterers including prolate and oblate spheroids in the resonance region, that is, for scatterers whose dimensions are comparable to the wavelength of light. Latimer and Barber (1978) made use of the extended boundary condition method (which is, of course, exact) to assess the worth of several approximations to solutions of the scattering from ellipsoids.

5.3 Measurement of Polarized Light

The great practical disadvantage of the Stokes vector is that its four parameters are not individually measurable by a physical system. The light to be determined is passed through a series of polarizing elements; there must be four independent measurements to allow deduction of the four Stokes parameters.

Most textbooks on light give an analysis table for the estimation of the state of polarization of a light beam. Visual examination through a linear polarizer, and a combination of a quarter-wave retarder and the polarizer, allows the observer to deduce the presence or absence of linear, elliptical or circular polarization.

This elementary procedure can be automated by arranging for a quarter-wave plate and a polarizer to be rotated in front of a photocell. One such system has been used for some years in the observation of noctilucent clouds (Gadsden 1975). The experimental details are as follows.

A small telescope has a rectangular aperture in its focal plane, some 60×3 arc minutes subtense. This aperture is manually rotatable and allows the field of view of the telescope to be placed *along* a noctilucent cloud band or stripe. Behind the rectangular aperture, there is a field lens imaging the telescope objective onto the cathode of a photomultiplier.

Between the field lens and the photomultiplier there is a quarter-wave retarder and a polaroid screen. Each of these may be rotated, in the opposite directions to one another, and such an arrangement was used at first. However, there is an advantage in simply having the retarder rotating and the polaroid fixed. The light incident on the photocathode thereby is linearly polarized in a fixed direction; any cathode sensitivity to amount, direction or type of polarization becomes of no importance.

The retarder may not be exactly a quarter-wave plate for all the wavelengths to be used. The polaroid is not an ideal polarizer. In general, the F matrix for the retarder will need to be written;

$$F_1 = \begin{vmatrix} 1 & 0 & 0 & 0 \\ 0 & \cos^2\frac{\delta}{2}+\sin^2\frac{\delta}{2}\cos4\beta & \sin^2\frac{\delta}{2}\sin4\beta & -\sin\delta\,\sin2\beta \\ 0 & \sin^2\frac{\delta}{2}\sin4\beta & \cos^2\frac{\delta}{2}-\sin^2\frac{\delta}{2}\cos4\beta & \sin\delta\,\cos2\beta \\ 0 & \sin\delta\,\sin2\beta & -\sin\delta\,\cos2\beta & \cos\delta \end{vmatrix} , \qquad (5.8)$$

where β = the angle of rotation of the retarder, which has a retardance equal to $(\delta/2\pi)$ wavelengths.

If we write k_1, k_2 for the principal transmittances of the polaroid, and set the axis of the polaroid vertical, the F matrix for it will be

$$F_2 = 0.5 \begin{vmatrix} k_1+k_2 & k_1-k_2 & 0 & 0 \\ k_1-k_2 & k_1+k_2 & 0 & 0 \\ 0 & 0 & 2\sqrt{k_1k_2} & 0 \\ 0 & 0 & 0 & 2\sqrt{k_1k_2} \end{vmatrix} . \qquad (5.9)$$

The light that passes through retarder and polaroid to the photocathode has the Stokes vector

$$\{I,M,C,S\} = F_1\,F_2\,\{I_0,M_0,C_0,S_0\}, \qquad (5.10)$$

and if we write B, the photocell signal, to be proportional to the intensity, I, of the light transmitted by the two polarizing elements, then

$$B = 0.5(k_1+k_2)I_0 + 0.5(k_1-k_2)\left\{ M_0\cos^2\frac{\delta}{2} + (M_0\cos4\beta \right.$$
$$\left. + C_0\sin4\beta)\sin^2\frac{\delta}{2} - S_0\sin\delta\,\sin2\beta \right\} \qquad (5.11)$$

If the polarizing elements are *ideal*, that is, if $k_1 = 1$, $k_2 = 0$, $\delta = \pi/2$, the photocell signal can be written:

$$B = 0.5(I_0+0.5M_0) + 0.25(M_0\cos4\beta + C_0\sin4\beta) - 0.5S_0\sin2\beta. \qquad (5.12)$$

In either case, (5.11) or (5.12), linearly polarized light gives a 4β variation only and circularly polarized light a 2β variation only. This forms a simple test for the presence of a small amount of circular polarization in the presence of a larger amount of linear polarization. Alternate extrema of the 4β variation have larger amplitudes than the intervening extrema. This is shown clearly in Fig. 5.2, which is part of a recording made during observations of noctilucent clouds on June 28–29, 1984.

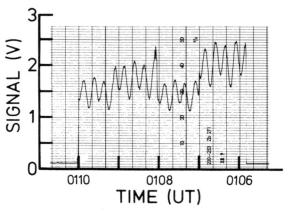

Fig. 5.2. Polarimeter recording of light scattered from a noctilucent cloud observed on June 29, 1984 at 0106 UT. The effective wavelength of observation is 440 nm and the angle of scattering is 47°. The field of view was centred at 11.5° above the horizon, azimuth 60°. Two strips of the cloud were observed (change is field of view at 0107 and 0108 UT) and the difference in radiance is clear. The dark signal at this time was 0.12 V and is shown on the record up to 0105:40 UT and after 0110 UT

Estimation of $\{I_0, M_0, C_0, S_0\}$ can be done by analyzing the photomultiplier signal for the D.C. level and the 2β and the 4β variations. An alternative, which has been used for several years, is to record the signal as integrals over successive 22.5° intervals of the retarder rotation and to solve four sets of four simultaneous equations set up in the form:

$$
\begin{aligned}
a_{11}I + a_{12}M + a_{13}C + a_{14}S &= B_1; \\
a_{21}I + a_{22}M + a_{23}C + a_{24}S &= B_2; \\
a_{31}I + a_{32}M + a_{33}C + a_{34}S &= B_3; \\
a_{41}I + a_{42}M + a_{43}C + a_{44}S &= B_4.
\end{aligned}
\qquad (5.13)
$$

The coefficients a_{ij} are estimated through calibration of the entire system. The telescope is pointed at a uniformly bright screen and a large polaroid and quarter-wave retarder are placed over the telescope objective. By rotation of the quarter-wave retarder in front of the objective, a series of B_i sequences is recorded for four distinct states of polarization. These are chosen to be the three orientations of linearly polarized light (vertical, horizontal and 45° in between) together with circularly polarized light.

Thus, the sets of the B_i refer to

$$
\begin{aligned}
B_1&: C_0 = S_0 = 0; M_0 = I_0; \\
B_2&: C_0 = S_0 = 0; M_0 = -I_0; \\
B_3&: M_0 = S_0 = 0; C_0 = I_0; \\
B_4&: M_0 = C_0 = 0; S_0 = I_0.
\end{aligned}
\qquad (5.14)
$$

The polarizer and retarder used in the calibration are not *ideal* polarizing elements although they are good. Use of a second (fixed) polaroid over the objective permits the assessment of their actual polarizing properties. These experimentally determined properties are used to adjust the above calibration (which was based on assuming ideal polarizers) to take the less than ideal properties into account. Finally, the state of polarization of the illuminated screen is estimated using the almost calibrated telescope and allowance is made for the slight degree of polarization that is almost certainly present; an iteration is practical and leads quickly to self-consistency in the calibration process.

5.4 Polarization Measured from Ground Level

Witt (1960) used a 35-mm stereoscopic camera which had two polarizing analyzers in front of the two objective lenses. A rotating frame allowed the operator to set the planes of the analyzers (held always at 90° to another) to lie in, and perpendicular to, the previously calculated direction of the plane of scattering. He had two pairs of filters, with effective wavelengths of 490 and 610 nm that could be placed over the objectives of the camera. Photographs were obtained on the bright display of Aug. 10, 1958 and 11 pairs of exposures with the red filters and 7 pairs with the blue were used in the subsequent analysis. The brightness of the display was some two to five times greater than that of the surrounding sky and Witt therefore made no allowance for any effect of the foreground twilight. His measurements of the degree of polarization, plotted against scattering angle, show the expected monotonic rise when the scattering angle increased from slightly over 20° to just over 60°.

Figure 5.3 has been redrawn from Witt's paper; the lines show the degree of polarization calculated from Mie theory, assuming no polarization in the light incident upon the clouds. The calculations were made using complex refractive indices appropriate for ice particles at a temperature of 130 K, viz.

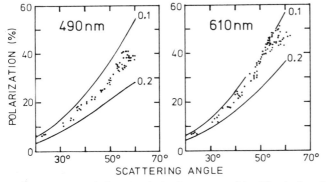

Fig. 5.3. Degree of (linear) polarization measured by Witt during observations of a noctilucent cloud on August 10-11, 1958. The *lines* plotted are the degree of polarization calculated from Mie scattering theory for ice spheres of radii equal to 0.1 and 0.2 μm, assuming that the cloud was illuminated by unpolarized light (Redrawn from Witt 1960)

$$\lambda = 490 \text{ nm:} \qquad m = 1.3071 - 2.8 \times 10^{-9} \text{ i;}$$
$$\lambda = 610 \text{ nm:} \qquad m = 1.3028 - 1.1 \times 10^{-8} \text{ i.}$$

Two radii were considered, 0.1 μm and 0.2 μm, and these are plotted in Fig. 5.3.

The measured polarization is consistent with scattering from dielectric spheres of radius 0.13 μm (490-nm filter) or 0.12 μm (610-nm filter).

Vasilyev (1962), Willmann (1962) and Tarasova (1962) reported measurements made in the USSR. Vasilyev's data extend over a rather limited range of scattering angles: between 15.5° and 20.3°. In this range, he found a degree of polarization between 0.07 and 0.23, with the plane of polarization lying in the scattering plane. Vasilyev suggested that the radius of the scatterers is 0.75 μm, with a column density of 10^6 m^{-2}.

This behaviour is supported by the observations reported by Tarasova who found that the plane of polarization lies close to the scattering plane, while the plane of polarization of the twilight sky (in the absence of noctilucent clouds) lies, as expected, perpendicular to the scattering plane. As she pointed out, the degree of polarization measured for the sum of twilight and noctilucent cloud scattering is necessarily lower than that for the noctilucent cloud alone, without a foreground of twilight.

Willmann (1962) presented in some detail data that show results similar to that of Witt and he suggested that particle radii (for a refractive index of 1.33) of 0.135 μm are present. Willmann's data are extensive; there are 382 measured points listed, with an estimated precision of 0.02 to 0.04 in the degree of polarization. Willmann's averages of the data, for intervals of 1 are plotted in Fig. 5.4. The lines are calculated from Mie scattering theory for two radii (0.1 and 0.2 μm) as for Fig. 5.3 but with the refractive index set to

$$m = 1.3042 - 6.3 \times 10^{-9} \text{ i.}$$

He used a photographic method with three fixed analyzers. He was able, therefore, to observe the direction of the plane of polarization and to compare it with the direction of the scattering plane.

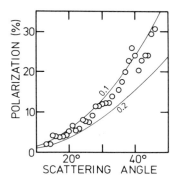

Fig. 5.4. The degree of polarization measured by Willmann (1962) during the night of July 30–31, 1959. The *plotted points* are averages in 1° intervals of scattering angle. The *lines* are the degree of polarization calculated from Mie scattering theory for ice spheres of radii equal to 0.1 and 0.2 μm, assuming that the cloud was illuminated by unpolarized light

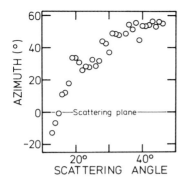

Fig. 5.5. The azimuth, θ, of the plane of polarization measured by Willmann (1962). The *plotted points* are averages in 1° intervals of scattering angle

Willmann found that the direction of the plane of polarization does not lie in the scattering plane, nor at right angles, but changes systematically with change in scattering angle. Willmann's data are plotted in Fig. 5.5 and the change is clearly much larger than the scatter of the averages of the measured points.

Let us examine the possible cause of such a rotation of the plane of polarization for different scattering angles. For illustration, consider the case of the scattering angle equal to 50°. It can be recalled [from Eq. (5.7)] that the scattering matrix for very small, spherical scatterers (Rayleigh scattering) contains only 6 non-zero coefficients out of the 16. Inserting the value $\theta = 50°$:

$$
F_{Ray} = \begin{matrix}
0.707 & -0.293 & 0 & 0 \\
-0.293 & 0.707 & 0 & 0 \\
0 & 0 & 0.643 & 0 \\
0 & 0 & 0 & 0.643
\end{matrix} \tag{5.15}
$$

Incident unpolarized light, $\{1,0,0,0\}$, will give scattered light that is partly (41.5%) plane-polarized perpendicular to the scattering plane, $\{0.707, -0.293,0,0\}$. If the incident light is plane-polarized, $\{1,1,0,0\}$, the scattered light will be plane-polarized in the same direction as the incident light, $\{0.414,0.414,0,0\}$. If the incident light is plane-polarized at 45° to the scattering plane, $\{1,0,1,0\}$, the scattered light is plane-polarized at a small angle (12°) to this direction, $\{0.707,-0.293,0.643,0\}$.

Note that for Rayleigh scattering, the incident light must contain some circular polarization for the scattered light to contain circular polarization; e.g. $\{1,0,0,1\}$ gives $\{0.707,-0.293,0,0.643\}$ which is fully elliptically polarized light, with 91% circular polarization and the major axis perpendicular to the scattering plane.

These calculations refer to scatterers which are much smaller than the wavelength of light with isotropic polarizability. If scatterers consisting of ice spheres at 140 K, and radii of 0.2 μm, are taken to be more representative of a noctilucent cloud, then

$$F_{mie} = \begin{vmatrix} 2.180 & 0.520 & 0 & 0 \\ -0.520 & 2.180 & 0 & 0 \\ 0 & 0 & 2.108 & 0.193 \\ 0 & 0 & -0.193 & 2.108 \end{vmatrix}. \qquad (5.16)$$

There are now two more coefficients appearing in the array, off the diagonal. The degree of polarization of the scattered light with unpolarized incident light is lower than with Rayleigh scattering (23.9% instead of 41.5%) and circular polarization is present when the incident light is obliquely plane-polarized. Incident light $\{1,0,1,0\}$ gives elliptical polarization in the scattered light, $\{2.180,-0.520,2.108,-0.193\}$; the direction of polarization is at a smaller angle (7°) to the 45° plane and there is 8.9% circular polarization present.

As we have seen, Willmann (1962) observed a change in azimuth of the plane of polarization with a change in scattering angle (Fig. 5.5). Is it possible that this comes about because there is partial linear polarization in the illumination of the noctilucent cloud? Such a state of affairs could come about by there being an appreciable contribution to the illumination on the cloud from sunlight scattered off the lower atmosphere. In this case, however, one would expect the measured polarization azimuths to show some symmetry about the direction of the solar meridian but there is no sign of this in the data.

All the measurements reported from the USSR have been analyzed with allowance for the foreground twilight. But those, and the measurements of Witt, make an initial tacit assumption that the scattered light shows only linear polarization. Gadsden (1977) reported a series of measurements of all four Stokes parameters.

The results for the degree of polarization are plotted against scattering angle in Fig. 5.6. In redrawing this graph, the distinction between different wavelengths has been removed. The line shows the degree of polarization expected for a spherical scatterer with a radius equal to a small fraction of a wavelength. The measurements have been corrected for the twilight foreground by measuring only the brighter features in a noctilucent cloud that have a dark region lying immediately next to the feature. The dark region is measured

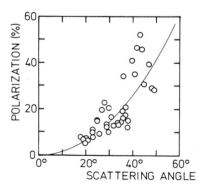

Fig. 5.6. The degree of polarization measured by Gadsden during 1974, 1975 and 1976. The *line* plotted behind the points is the degree of polarization expected for small spheres of ice with radii equal to $\lambda/63$ (Redrawn from Gadsden, 1977; the distinction between observations at 400, 450, 575 and 675 nm wavelengths has been removed)

immediately after the measurement of the bright region. Normally this involves comparing areas of sky only a few minutes of arc apart. In the case of a bright stripe, the fainter regions on each side of the stripe are measured to give foreground subtraction.

There is often a certain amount of "super-polarization" present, that is, the degree of polarization is greater than that expected for Rayleigh scattering.

The results for the direction of the major axis of the polarization ellipse (Fig. 5.7) do not verify Willmann's measurements. The direction is always within 45° of the perpendicular to the scattering plane although there appears to be a systematic, small anticlockwise rotation of direction away from perpendicularity (that is, anticlockwise on *both* sides of the solar meridian).

The results also show that there is a detectable amount of circular polarization in the scattered light and that there seems to be a systematic change in the proportion of circular polarization with change in wavelength. The red and blue parts of the spectrum (wavelengths 400, 450 and 675 nm) show typically 0.02 to 0.08 left-handed circular polarization, while scattered light at 575 nm shows 0.00 to 0.02 right-handed polarization.

With reference to the principles of symmetry that underlie the theory of the scattering, matrix F shows that circular polarization can be present in the scattered light as a result of any or all of three causes:

1. If the incident light is unpolarized and there is a preferred orientation for the (non-spherical) scatterers;
2. If the incident light contains some circularly polarized light;
3. If there is some light linearly polarized in a direction oblique to the scattering plane.

It is possible that the presence of circular polarization in the scattered light arises to a large extent from the presence of forward-scattered sunlight in the light incident upon the clouds. This would account for the absence, or small amount

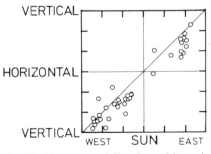

Fig. 5.7. The measured directions of the major axes of the polarization ellipses, from observations in 1974, 1975 and 1976 (Gadsden 1977). The *points* are plotted against the orientation of the scattering plane, which is vertical (90°) on the meridian of the Sun and increases towards 180° for points in the sky to the *east* of the Sun and decreases to 0° for points in the sky to the *west* of the Sun. Because the Sun is below the horizon, there is no part of the sky for which the scattering plane lies horizontally (0° or 180°)

of reversed handedness, of circular polarization in the results for the yellow (575 nm) filters compared with the data for the blue and the red filters. The 575-nm sunlight suffers marked absorption during passage through the ozone layer (as seen, for example, in the spectra shown in Fig.3.2). Sunlight that is scattered in the forward direction from the troposphere near the terminator may well contain a significant proportion of linearly or circularly polarized light. It is possible, therefore, that the sunlight incident upon the clouds in the mesosphere is unpolarized light only where an ozone absorption band stops this scattered light from reaching the clouds.

The polarization measurements reported by Gadsden (1977) were made with a single-channel polarimeter. The different wavelengths were selected by manual rotation of a circular wedge interference filter. The observations were thus inevitably made *sequentially* and the results for the different wavelengths do not refer to observations made on the same area of noctilucent cloud. Indeed, the results quoted above are from a set gathered over several years' observations and measurements at all three wavelengths were not necessarily obtained during any one night.

A more recent instrument consists of three identical telescopes, with interference filters over the objectives to give three channels each with a particular (fixed) wavelength. The wavelengths were chosen to lie (1) at the blue edge (440 nm) of the Chappuis absorption band; (2) directly in the middle (596 nm) of the Chappuis band; and (3) on the red edge (791 nm) of the band. The fields of view (54×3.3 minutes of arc) were identical (to within ± 3.5 arc minutes); the recordings (3.1-s integrations) were made simultaneously into three buffered counters which were read out once a minute into a 3×16 array. The Stokes parameters at three wavelengths can thus be obtained simultaneously from the same cloud feature.

Figure 5.8 shows some results obtained from using this instrument. The proportion (S/I) of circular polarization in the light scattered from noctilucent clouds is plotted for the three wavelengths separately. It is clear that in the blue, there can be a few percent of *left-handed* circular polarization; in the yellow, there is typically a few percent of *right-handed* circular polarization; and in the deep red, a small amount of right-handed circular polarization.

These data indicate that noctilucent clouds do indeed contain non-spherical scatterers, which are illuminated by partly plane-polarized light. In the blue, there is a large proportion of tropospherically scattered light incident on the clouds and this imposes left-handed circular polarization. In the yellow, light scattered from the troposphere along slant paths will suffer absorption in the stratospheric ozone layer and will therefore not be available to impose left-handed circular polarization. In the yellow, we are seeing the true right-handed circular polarization of scattering from noctilucent clouds. In the deep red, there is little scattering of light from the troposphere and a proportion of right-handed circular polarization is seen. This proportion is less than in the yellow because the wavelength of the light is greater and the scatterers are a smaller fraction of the wavelength of light.

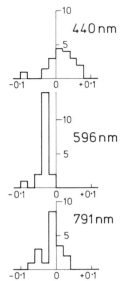

Fig. 5.8. Distributions for three wavelengths of measured proportions of circularly polarized light. Positive values are left-handed polarization, negative values right-handed

Again, if the occurrence of circularly polarized light in the light scattered from a noctilucent cloud was to arise principally from the presence of linearly or circularly polarized light in the illumination of the cloud, one might expect the amount of circular polarization to be rather uniform over large areas of a noctilucent cloud display. On the other hand, if the scattering properties of a particular part of a display cause the scattered light to contain circular polarization, the amount of circular polarization seen in the light from a display might well be concentrated in limited areas of the display.

To check this point, observations were made with a television camera which had an analyzer in front of its objective lens (Gadsden et al. 1979). The analyzer consisted of a polaroid filter fixed in position over the lens, and in front of this filter there was a rotating quarter-wave plate. A blue-green filter gave a certain amount of spectral selection. The output from the camera was recorded on tape for subsequent analysis in a video-integrator which allowed the signal from part of the picture to be selected and displayed on a chart recorder (Taylor 1981). This system has been called a "post-hoc photometer". Only one display was observed during the summer that this television system was operating, that of July 28, 1978. Examination of the video recording (Fig. 5.9) showed that on this occasion circularly polarized light was coming from one small area of the clouds, with neighbouring areas showing none.

The proportion of circular polarization was 0.005 ± 0.0013, which is small when compared to the earlier measurements of Gadsden (1977). The distinct localization of the area giving circular polarization argues against it having come about through tropospheric scattering contributing to the incident sunlight. The circular polarization is probably a result of the scattering process in the clouds

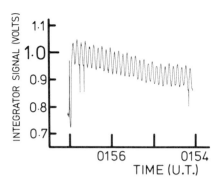

Fig. 5.9. Post-hoc photometry of a video recording; the alternating height of maxima is a clear indication of the presence of circular polarization in the light scattered from a noctilucent cloud (Gadsden et al. 1979)

themselves. More information is needed to elucidate the processes occurring; the interpretation in some ways shares the uncertainty of interpretation of the spectral radiance measurements.

5.5 Measurements of Polarization from Rockets

At first sight, it is a simple task to make in situ measurements from a sounding rocket passing through a noctilucent cloud. The facts of ballistics complicate things. Above the stratosphere, the flight of a sounding rocket is very nearly ballistic; the motors of the rockets that are used normally burn for only the first few kilometres (if that) of the flight. For the rest of the flight, the rocket is just a pointed cylindrical box of instruments and radio transmitter, with fins at one end. It will be spinning on its long axis at a few revolutions per second for stability. The rocket casing also processes, with a cone angle of perhaps 10° or 20°. The time actually spent passing through a cloud is short: if the apogee is 100 km, the rocket will pass through the whole interval from 80 to 90 km in less than 20 s. If the apogee is 150 km, the time is halved. On the way down, the rocket normally falls tail first; air drag is not enough to make it turn over until it returns the stratosphere.

It appears that Witt (1969) was the first to project a photoelectric polarimeter through a noctilucent cloud. There were problems with illumination of the polarimeters by the Sun but data were successfully obtained at two wavelengths (366 nm on one channel and 534 nm on the other). The scattering angle was 86°. The noctilucent cloud was entered at 82.4 km altitude, and the rocket seems to have broken clear of the layer by about 84.4 km. There was a high degree of linear polarization, and if the cloud particles were scattering as monodisperse dielectric spheres of refractive index 1.33, an upper limit of 0.2 μm in radius was indicated.

Witt was involved in another rocket sounding from Kiruna (69°N) the following year (Witt et al. 1976). For this flight, the wavelengths chosen for the polarimeters were 256 and 536 nm. Both instruments sensed a scattering layer at altitudes between 85.5 and 89 km on both ascent and descent. The 256-nm polarimeter signal was affected to a considerable extent by fluorescence from upper atmosphere nitric oxide. Witt et al. deduced an upper limit to the particle radius of 0.05 μm. These may well be the particles available for nucleation of noctilucent clouds when the temperature or humidity is suitable.

Witt et al. (1971) reported on two flights from Kiruna in 1971 using more ambitious polarimeters. These involved measurements at seven wavelengths (214, 309, 366, 453, 536, 589 and 762 nm) and there were a number of other instruments (airglow infrared photometers, nitric oxide ionization chamber, solar irradiance photometer, ion collector and Faraday rotation). The preliminary results from the polarimeters showed noctilucent clouds on both flights. On the July 31 flight, the layer was between 82 and 83 km; the following night, the cloud layer was higher, at 85–87 km. At 453 nm, the degree of polarization was a little higher than that expected from molecular scattering.

As part of a campaign to launch particle collectors through noctilucent clouds (see Chap. 6), Tozer and Beeson (1974) obtained polarimetric data from four flights over Kiruna. The instrument on each of the flights contained two channels sharing a common entrance pupil by the use of a split fibre optic bundle. Filters gave effective wavelengths of 540 and 410 nm. The polarimetric data of Tozer and Beeson are plotted in Fig. 5.10. There was no penetration of a noctilucent cloud on the second flight, Aug. 1, 1971, so the data came from three flights, ascent and descent on two, ascent only on one.

At the large scattering angles involved, the degree of polarization is close to 1.0 until the scatterer radius becomes greater than approximately 0.1 μm. The Tozer and Beeson results indicate that scatterers as large as 0.2 μm are most unlikely to be present. However, it should be recalled again that the calculations giving the reference curves in Fig. 5.10 assume that the incident light is unpolarized. The rocket flights occurred during the middle of summer nights at Kiruna, in northern Sweden, so the illumination conditions are similar to those applying to ground level observation of noctilucent clouds.

Heintzenberg et al. (1978) discussed the extraction of the median radius, the largest radius and the width of the distribution of particle sizes in a cloud from inversion of observations of the Stokes parameters I and M. They reported observations made at four wavelengths, 214, 366, 453 and 536 nm. Scattering angles, θ, for observations from "our recent rocket launchings from northern

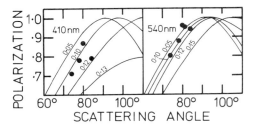

Fig. 5.10. The polarization measurements of Tozer and Beeson (1974) made with rocket-borne photometers. The degree of polarization is plotted against scattering angle for two wavelengths (410 and 540 nm) and the *lines* are values calculated from Mie scattering theory for ice crystals of radii close to 0.1 μm

Sweden" lay between 70° and 90°. Data are plotted for one flight and show a degree of polarization at 214 nm lying between 0.55 at $\theta = 70°$, and 0.65–0.89 for $\theta = 90°$. There was, however, apparently some optical interference present in this channel and the authors stated that the measured degree of polarization is to be regarded as a lower limit. The results from the three longer wavelengths show the measurements clustering closely together, running from about 0.74 ($\theta = 70°$) to 0.90–0.97 for $\theta = 90°$. Figure 5.10 shows that these degrees of polarization indicate typical radii around 0.1 μm. The analytical inversion technique used by Heintzenberg et al. (1978) gives a maximum radius of 0.075 μm with the most likely radius lying in the range 0.06 to 0.075 μm.

5.6 Conclusions About Cloud Particle Sizes

In a similar way to the caution expressed in Section 4.6, but for different reasons, there remains uncertainty about the conclusions to be drawn from the polarization data. If the irradiation on noctilucent clouds is unpolarized, the presence of polarization directions slant to the scattering plane and the presence of some degree of circular polarization suggest that there are cloud particles large enough (radii > 0.2 μm), or appreciably non-spherical or anisotropic, in the cloud. The *amount* of polarization that is measured points unequivocally to there being no cloud particles larger than approximately 0.1 μm in radius.

On the other hand, if the irradiation of the clouds is assumed to be partly polarized, evidence of rotation of the polarization plane and of circular polarization may no longer be used to infer the presence of the larger particles. Equally so, in this case, the high degree of polarization is no longer immediately interpretable in terms of the *absence* of the larger particles.

Bohren (1983) considered these problems and concluded that because he could not find a mechanism to orient larger scatterers (> 0.1 μm radius), that the circular polarization that has been measured must arise from there being some proportion of polarization in the light incident on the noctilucent cloud. Bohren thus accentuated the effect of multiple scattering on the irradiance of the cloud layer and rejected the alternative interpretation of larger, non-spherical scatterers. Gadsden (1983) to some extent justified the deduction of scatterers with radii greater than 0.1 μm by invoking Brownian rotation of non-spherical scatterers with observational selection to provide an apparent preference in the axial direction of the scatterers.

Perhaps the conclusion to be drawn at this stage concerning the characteristic size of the scatterers, larger or smaller, is to be expressed in the two words that can be the verdict of a jury in Scotland: "Not Proven!"

6 Rocket-Borne Sampling

6.1 Introduction

As seen in preceding chapters, the optical data have given conflicting, and to some extent contradictory, information on the characteristic size of the particles making up a noctilucent cloud. Given this situation, the answer would seem to be to go up to a noctilucent cloud and collect some of the particles for laboratory examination.

This is a difficult technical problem, however. The rocket vehicle is likely to be travelling supersonically through the cloud and there will be shocks around the casing and spreading from the nose cone. Simply collecting the cloud particles on trays may therefore be difficult because the airflow may stop the particles from reaching the trays. Local turbulence may cause the particles to spill out of the trays.

For aerodynamic stability, the collectors cannot be exposed until the rocket is in the upper atmosphere. Sometimes the collectors are deployed on arms extending from the casing in order to place them beyond the turbulence and shocks near the skin of the casing; air drag on the collectors when deployed in the lower atmosphere would pose unacceptable mechanical problems. Thus, collectors must be *deployed* or *uncovered* at the right time (height) and *retracted* or *covered* in time for the descent through the lower atmosphere.

In addition, the collectors must be recovered; either a sealed package has to be parachuted separately to a soft landing down range from the rocket launch or the spent rocket casing has to be found and the collector package recovered.

The principal remaining difficulty, after all the flight problems, is in identifying those particles which have been collected from the noctilucent cloud. There will inevitably be contaminant particles (appearing similar to the cloud particles) which have lodged in the collectors before, during or after the flight.

The problems have been overcome by several experimenters and the results of their successful flights are discussed in this chapter. Overcoming the problems has led to several ingenious techniques for tagging the particles that we are interested in. Describing these techniques must also be included in the discussion of the results just as the general discussion of polarized light was necessary for interpreting the optical data.

6.2 Flights over Sweden in 1962 and 1967

Two successful flights, on Aug. 7, 1962 (no noctilucent cloud) and Aug. 11, 1962 (noctilucent cloud), were launched from Kronogard (66°N) in northern Sweden (Hemenway et al. 1964b).

Collecting surfaces were exposed between 75 and 98 km altitudes during the ascent of each rocket. The second flight, through a noctilucent cloud over the launch site, showed a column abundance of particles at least 1000 times greater than on the first flight which was made in the absence of clouds.

Each rocket carried a variety of particle collectors and the particles were analyzed (after post-flight recovery) by electron diffraction, neutron activation and electron beam techniques. Some evidence was found for water in the collected specimens; a significant fraction of the particles showed the presence of a volatile coating.

An electron microscope photograph is reproduced in Fig. 6.1. This shows a particle collected on an aluminium-shadowed nitrocellulose film, with extra chromium shadowing at near-grazing incidence (5°). There is a small central

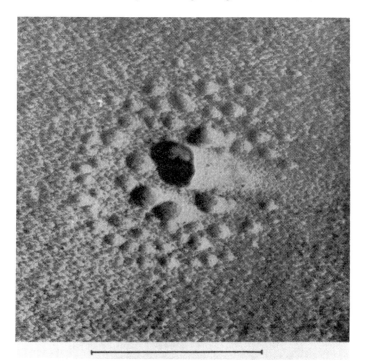

Fig. 6.1. An electron microscope photograph of a noctilucent cloud particle collected on a nitrocellulose film during a rocket flight over Kronogard (Sweden) on 11 August 1962. The *bar scale* is 1 μm. This and the photograph in Fig. 6.2 are reproduced by kind permission of Tellus magazine

particle, approximately 0.15 μm in diameter, surrounded by an area of nitrocellulose film which has been puckered over an area more than 1 μm in
diameter. There are small bumps in the film; the shadowing produces the same
pattern of light and dark that the central, presumably solid, particle gives. The
picture certainly gives one the impression that there was a layer of material
surrounding the central particle which has evaporated since collection on the
film. The obvious candidate for the volatile material is ice.

Some particles had no central core and appear to have consisted of volatile
material only (Fig. 6.2). Several dozen such structures were found in the cloud
samples and none was found on either the (shielded) control surfaces or on
surfaces exposed during a rocket flight with no noctilucent cloud present. The
structure shown in Fig. 6.2 is 9 μm in diameter but this is not necessarily the
diameter of the original cloud particle. Melting an ice sphere with a diameter
of 7.4 μm will give a hemispherical drop of water that is 9 μm in diameter.
Furthermore, the cloud particle may have melted while the collecting film was
still exposed and the water droplet then flattened, spread out or was shaken
across the film by the air blast of the rocket flight.

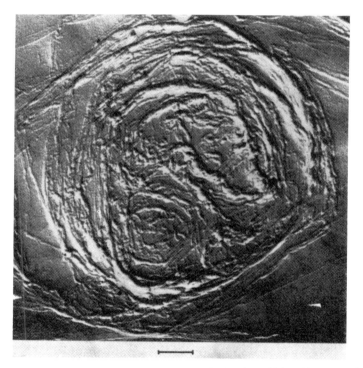

Fig. 6.2. An electron microscope photograph of a nitrocellulose film exposed during a rocket flight
through a noctilucent cloud (see legend to Fig. 6.1). In this case, the film appears to have been
wrinkled by the melting of a volatile particle. The *bar* scale is 1 μm

Hemenway et al. summarized the findings from the collection of experiments as follows:

1. The nuclei of noctilucent cloud particles are of extraterrestrial origin.
2. These nuclei have an integral size distribution of the form $N = Ad^{-p}$; $3 < p < 4$.
3. The size distribution of nuclei cuts off sharply at about 0.05μ/diameter.
4. A significant fraction of the particles were ice-coated when collected.
5. The layer concentration in the sampled layer (75–98 km) is at least 1000 times greater in the presence of clouds than when they are not seen.
6. The particle concentration in a vertical column through the cloud display of 11 August 1962 was greater than 8×10^{10} particles m^{-2}. (This figure is rather larger than most of the estimates quoted in Table 4.3).
7. If the particle density (number per unit volume) decreased exponentially with height, then the scale height is about 2 km.

The size distribution of 93 coated particles is shown in Fig. 6.3, where the distribution of radii of the complete particle (nucleus + halo) is shown separately from that of the radii of the nuclei. In addition, there seems to be a relation between the size of the coated particles and the size of the solid nucleus seen inside many of them. The larger nuclei appeared to have attracted thicker coats of volatile material.

Two different preparations of nitrocellulose collectors were used. Both had 20-nm-thick films for substrate; one type had a thin overcoat of aluminium, the other was coated with a layer of fuschin dye (Soberman et al. 1964). The fuschin, deposited in a vacuum system, was to establish the presence of a liquid coating (or a solid coating which subsequently melted) on collected particles. The thin film of dye would be disrupted by droplets of liquid.

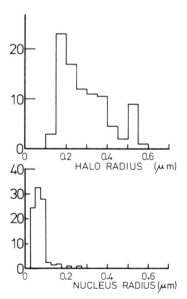

Fig. 6.3. Histograms of particle radii based on 93 particles found on a nitrocellulose film flown through a noctilucent cloud (Hemenway et al. 1964a). The *upper* plot is the radii of particles showing haloes; the *lower* plot is the radii of their solid nuclei

Hemenway et al. (1964a) discussed the samples obtained with these two films in some detail; they presented cumulative distributions of the particle sizes found on each of the detectors. Their data have been replotted in Fig. 6.4 as size distributions rather than cumulative plots.

Calcium film collectors were also used (Linscott et al. 1964). These consisted of a package of evaporated layers on a lucite substrate. From the lucite outwards, there was first a layer of lithium fluoride approximately 100 nm thick, then a layer of aluminium (5 nm) and a relatively thick layer of calcium which was in turn covered by 5 nm of aluminium and a layer of paraffin. The whole sandwich was topped with a layer of silicone oil. This seals the surface against water vapour and also acts as a trap for impacting particles.

The detector was judged, after tests, to be capable of detecting ice crystals at least down to 0.25 μm radius; metallic and abrasive particles 0.1 μm and upwards in size were fired with speeds of around 800 m s^{-1} at the layer in tests and only those particles larger than 10 μm were found to rupture the calcium film.

Results were obtained from four such surfaces, two on each flight. The surfaces from the second flight (in the presence of a noctilucent cloud) were clearly different from both the controls and from the surfaces exposed on the first flight. This shows up clearly in the pictures reproduced in Fig. 6.5. The (unexposed) control surface on the left is much less damaged than the exposed surface shown on the right. The pictures were taken of the same sample surface;

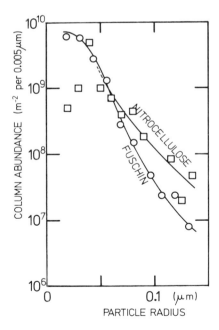

Fig. 6.4. Logarithmic plots of the size distributions found from two different collectors flown through a noctilucent cloud (Hemenway et al. 1964a)

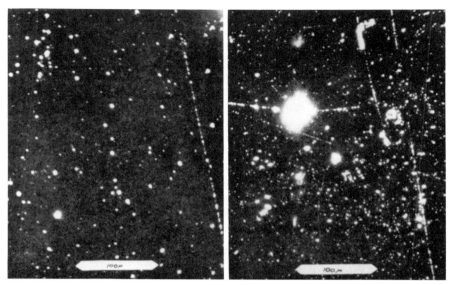

Fig. 6.5. Micrographs of two areas of a calcium film detector flown through a noctilucent cloud on 11 August 1968. The area shown on the *left* was shielded from particle impact; the area on the *right* was exposed. The *scale bar* denotes 100 μm. The photographs are reproduced by kind permission of Tellus magazine

the only difference is that a shield was used to cover that part of the surface which is shown on the left.

There was a variable density of the calcium film and examination under phase contrast illumination showed a coarse texture to the exposed film. The film itself appeared to have become rather granular. In the centre of many of the grains there were approximately 4.5×10^9 m^{-2} small holes. This is not greatly at odds with the number of haloed particles seen on the collectors scanned under an electron microscope.

Laboratory tests using a radio-tagged water spray at low temperature and pressure suggested that the slight changes in reflection and transmission arose from approximately 1.5×10^{-5} kg m^{-2} moisture. Each cloud particle thus has 3×10^{-15} kg water associated with it on average (corresponding to a sphere of radius 0.9 μm).

Witt (1969) tried to repeat the observation of haloed particles on a flight in 1967, which was launched through a noctilucent cloud. The collector was a collodion film shadowed with palladium for electron microscope examination. The limit of detection was particles of radius less than 5 nm, yet only two particles were found on 4 mm^2 of the film. (This corresponds to a column abundance of $\sim 5 \times 10^5$ particles m^{-2}, a very much smaller figure than was expected.)

6.3 Flights over Sweden in 1970 and 1971

Hemenway et al. (1972) flew rockets from Kiruna (68° N) in northern Sweden during the summers of 1970 and 1971. The payloads included "Pandora" collectors described by Hemenway and Hallgren (1970). These collectors had the capability of inflight shadowing of films for subsequent electron microscopy. Their description of the collectors is repeated here:

The Pandora II collector . . . consists of eight collection arms, each approximately 3.7 cm wide and 15 cm long. Each collection arm contains a collection box approximately 7.3 cm long, 2.2 cm wide and 1.0 cm deep. . . . Most parts of the collector were made of anodized aluminum. Silicon rubber gaskets were used. All metal parts of the collector were individually ultrasonically cleaned, then assembled in our clean laboratory.
The in-flight shadowing consisted of two, five-mil tungsten wires within plexiglas shields, each on opposite sides of a set of eight electron microscope grids. Maxtaform H-2 grids were used to permit identification of the location of each collected particle on a coordinate system. Each tungsten wire had four beads of gold carefully located in the middle of the aperture of each slit in the plexiglass [sic] shields. The gold beads each weighed approximately 0.12 mg and were fused in place. The tungsten wires were kept under tension both to minimize distortion during the acceleration of the boost and to withstand the opening shock of the collector arms. In-flight shadowings were carried out by passing a sufficient current through the tungsten wire to evaporate the gold and break the electrical continuity of the wires in approximately 150 msec. The shadows were cast in pairs: the first initiated 0.3 sec after the collector arms were deployed and the second 0.3 sec before the arm retraction began. Reliability of the in-flight shadowing technique was checked by seeding each grid of each altitude interval of each flight with the aid of easily identifiable particles which were, therefore, doubly shadowed. No movement of these submicron test particles was detected.
The eight Pandora II collector arms were deployed sequentially, in pairs, in order to sample four altitude intervals between 75 and 150 km.

Much of the doubt about the origin of the particles is removed by the double shadowing during deployment of the collectors. There is a clear signature of two shadows to indicate particles that were acquired before the interval between two evaporations using separated sources of the shadowing material. Again, particles showing two components were collected. There was typically an electron-opaque core surrounded by low-density material in a rounded coating. In the 1970 collections roughly 70% of the particles were of this type, compared with but 15% from flights the following year. In both years the remaining particles were mostly single, submicron, irregular particles of high electron opacity.

In a later paper, Hallgren et al. (1973a) discussed the 1970 flights and presented the data in more detail. The haloed particles accounted for 85% of the total collected and were clearly round or elliptical, dropletlike particles of low electron density with varying amounts of high density material in their centre. The particles did not appear to be the same type as were seen in the 1962 flights but the authors reported that a difference in shadowing technique may account for this. The concentration of particles was about 10 times greater than was seen on similar flights from mid-latitudes (at White Sands, 32°N, in the USA) and 100 times less than was seen on the noctilucent cloud penetrations in 1962.

Hallgren et al. (1973b) reported on the 1971 flights carrying Pandora collectors. There were two flights, on July 31 and August 1; the first successfully penetrated a cloud, the second missed. In these collections, X-ray emission analysis was applied to the particles; the sensitivity of the system was such as to allow analysis of particles with radii not less than 0.26 m. Unless an element was present as a major fraction of the particle mass, it was unlikely to be detected. The results show that a large proportion of the collected particles contained elements of high atomic number (Hf, Ni, Co, Ce, etc.). The authors suggested that these particles must be of solar origin.

6.4 Flights over Canada in 1968 and 1970

Farlow et al. (1970; see also Farlow and Ferry 1972) have flown a different type of collector through a noctilucent cloud. Long arms (1.5 m) were folded inside the rocket casing and were arranged to open out at a preset time (and thus a preset altitude). Each arm carried pans (see Fig. 6.6) which were sealed under vacuum before being loaded into the rocket. After the required period for sampling, the arms retract and (as the rocket falls back into the lower atmosphere) the pans are pressed by air pressure into covers with gaskets which reseal the collection. In Fig. 6.6, a special collector with ribbons on the frame is shown flipped out beyond the main pans. This collector was used to obtain a high collection efficiency with submicron particles.

In the flight of 1968, launched from Fort Churchill (59°N) in northern Canada, the arms bearing the electron microscope screens began to open at

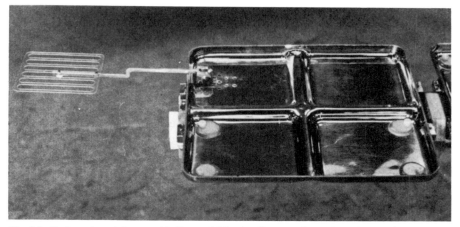

Fig. 6.6. Outboard module pan with flip-out (ribbon) collector as flown through a noctilucent cloud on 1 August 1968. The photograph is reproduced by kind permission of the American Geophysical Union

55-km altitude (rather lower than was planned) and damage to the screens was experienced. Farlow and Ferry (1972) were not sure that the collection was from a noctilucent cloud but they did see a considerable increase in numbers of collected particles over and above what they expected from flights at mid-latitude.

Their solid-particle collector was scanned first at a magnification of 4000 (see Fig. 6.7). The collecting surfaces were shadowed by a vacuum deposition of chromium from different directions before and after the flight. As we have seen, this means that particles with two shadows must have been on the collectors before the flight; particles with one shadow are later arrivals. Figure 6.7 shows six distinct types of particles. Types 1,3 and 6 are "obvious contaminants" because they are found in low numbers on all surfaces, exposed and unexposed. Type 4 seemed to be more frequent on the exposed than on the unexposed surfaces but was later eliminated by Farlow et al. from what they considered to be the genuine cloud particles. That leaves types 2 and 5 but Ferry and Farlow

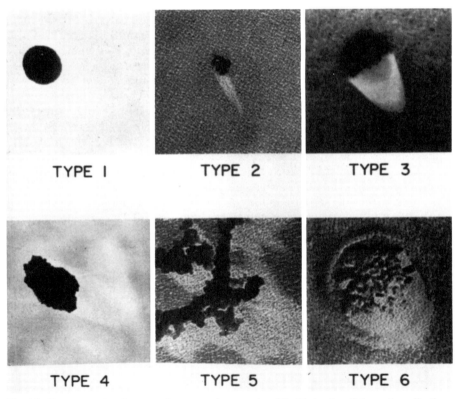

Fig. 6.7. Representative electron microscope photographs of the kinds of particles on the collecting surface of the 1 August 1968 flight. The photograph is reproduced by kind permission of the American Geophysical Union

(1972) later reported that collection through the noctilucent cloud of 1 August 1968 showed only type 2 particles.

Figure 6.8 shows the size distribution of these particles obtained from the data plotted in Farlow et al. (1970) together with those listed in Ferry and Farlow (1972). It is possible that the absence of particles of radius less than 0.025 μm, and the decrease shown in the number of particles of radii less than 0.05 μm, arise from the difficulty of detecting and identifying particles of this size, even though the magnification was increased to 15 000 in order to make the counts upon which Fig. 6.8 is based.

Water-droplet collecting surfaces were flown but showed no impacted particles. These surfaces were constructed from 50-nm-thick layers of polyvinyl alcohol placed onto a 50-nm film of nitrocellulose. The alcohol is insensitive to humidity but highly soluble in water. A droplet, or ice crystal which subsequently melts, dissolves part of the film and leaves the dissolved material at the edge of the drop during evaporation of the water, causing a raised rim and crater in the film.

Farlow and Ferry (1972) concluded that the type-2 particles that they collected on the 1968 flight were indeed noctilucent cloud particles and that they had been uplifted to mesospheric levels from the lower atmosphere. That is to say they questioned the cosmic origin of the cloud particles or nuclei.

This is because they found that large concentrations of particles ("dust") are present in the upper atmosphere only occasionally. This rules out, according to these authors, a steady-state deposition of solid material from the flux of sporadic meteors. The concentrations which they measured should result rather from the very slow settling rate of the dust in the upper stratosphere which allows a buildup of the density of particles. This buildup provides the *source* for locally high concentrations of dust at mesospheric heights by the action of convection ("uplifting"). It must be added that such localized upwelling with the speeds necessary to carry the dust particles aloft has yet to be observed.

Two further flights were made, also from Fort Churchill, in the spring of 1970, before the start of the noctilucent cloud season. The purpose of these flights was to measure the cosmic dust at the same latitude as the noctilucent

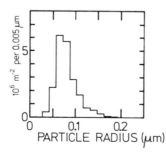

Fig. 6.8. Size distribution of so-called type 2 particles found on collectors flown through a noctilucent cloud (Farlow et al. 1970; Ferry and Farlow 1972)

cloud flight but before any uplift of low level particles had occurred. They found only a few dozen particles on the collecting surface where before they had several thousand.

6.5 Collectors Flown by Max-Planck-Institut Researchers, 1968 to 1971

Collectors and detectors of a rather different type to those described above have been flown on a series of rockets launched into noctilucent clouds by a group at the Max-Planck-Institut in Germany. There were flights in 1968 from Fort Churchill (Fechtig and Feuerstein 1970) and in 1969 from Kiruna (Fechtig et al. 1971). The following year (Rauser and Fechtig 1972), more data were obtained from a flight through a weak noctilucent cloud and in 1971 there were flights from White Sands (32°N) in April and Kiruna (68°N) on July 31 and August 1. Neither of the last two flights apparently penetrated a noctilucent cloud although both were made "in the presence of" clouds.

The dust collectors used on the flights had clean and highly polished silver, aluminium and copper surfaces, some of which were shielded by a grid, mounted some millimetres in front of the actual surface. The grid supported a 0.2-μm nitrocellulose film.

A second type of particle detector consisted of a concave hemisphere of tungsten acting as a grounded anode. Charge is released on the impact of a particle and the rise time and the amplitude of the pulse measured at a central cathode (−50 V) could be used to estimate both the speed and the mass of the particle.

The collector from the 1970 flight which penetrated a noctilucent cloud showed 290 penetration holes (diameters ranging from 1 μm to 50 μm) on a collector area of 1.62×10^{-4} m^{-2}. (This corresponds to 1.8×10^6 m^{-2} particles of radii greater than 1.0 μm.) Electron microscopy of an area equal to 25.4 mm^2 showed five accumulations of small particles ranging in radius from 0.022 to 0.1 μm, corresponding to a total of 2×10^5 m^{-2}. The collectors and detectors were exposed at 61-km altitude on the ascent of the flight, through apogee of 114.3 km, and closure took place at 99.5 km on the descent. Some ring structures (with radii up to 15 μm) both with and without central particles were noted. These did not appear on the control surfaces and Rauser and Fechtig interpreted these as

. . . probably produced by the impact of droplets or dirty ice crystals during flight. Vibration tests carried out on the collector films showed no holes comparable with those found on the flight samples.

Rauser and Fechtig (1973) summarized their findings as follows:

. . . particles in the micrometre and submicrometre range which are entering the atmosphere are decelerated to very low velocity (0.5–1 km s^{-1}) by the time they reach about 100 km altitude. However, the velocity profile between 70 and 100 km altitudes is very different from earlier model calculations. An explanation of the observed velocity profile is that particles falling through the cold summer mesopause are able to agglomerate or absorb condensable constituents like H_2O, CO_2 and related clusters as found in these altitudes . . . The existence of noctilucent clouds, therefore, may partly be explained by the presence of those particles in the region of minimum temperature which would correspond to the maximum of their mass x density phase.

6.6 Conclusions About Cloud Particle Sizes

The evidence from in situ collection is thus largely at odds with much of that from optical investigations. In Chapters 4 and 5, many opinions are cited which assert that there can be no cloud scatterers with radii larger than 0.05 to 0.1 μm. Other investigators suggest that the upper limit should be 0.3 μm or perhaps as large as 0.7 μm. We shall see later that modelling a noctilucent cloud, using good estimates of the local atmospheric conditions, supports the view that a cloud particle as large as 0.1 μm radius is rare. But the results of particle collections indicate a significant and important number of larger particles.

If these experimental results are not to be dismissed out of hand as spurious, or contaminated to a degree as to render the data useless, we are forced to accept the larger-particle optical results and to view with suspicion the indications from the modelling calculations.

7 Variation of Occurrence

7.1 Introduction

The theory of noctilucent clouds must first of all explain their diurnal and monthly appearance, then also the special effects which have been noted in past observations (cf. Schröder 1966b, 1968c). That is the sporadic appearance of noctilucent clouds, their absence during the months September-April in the northern hemisphere; the nearly constant height of about 82 km as well as the morphological picture (Figs. 1.1–1.7 and 3.9).

That explanation is only partly successful, but may be due, on the other hand, to insufficient knowledge of the various layers of altitude and, on the other hand, to the scarcity of general mathematical-physical estimates. Generally spoken Ertel (1938) pointed out that "an atmospheric apparition is considered as 'explained' when we have succeeded in deducing it on the basis of accepted principles of physics". With this assumption, the hypothetical character of past evaluations inclines one to replace these speculations by a more general treatment, which, to be sure, will still be subject to certain limitations. In doing this we will not obtain a clearly proven theorem but an explanation which may eliminate many of the uncertainties. It would be desirable to have increasing worldwide observations of this phenomenon, because the individual results can be checked only with extensive and reliable observational material.

7.2 Sunspot Cycle

A relationship between noctilucent cloud activity and reduced solar activity was suggested by Vestine (1934). He concluded that the relation with low sunspot activity was not satisfactory and could be fortuitous, because the degree of interest in these phenomena has varied widely over the years since 1885. Fogle (1966) analyzed data from the years 1885–1964 and found no correlation between the solar cycle and the frequency of occurrence of noctilucent clouds.

To study the relationship between the occurrence of noctilucent clouds and the level of solar activity, it is particularly important to know the reliability of the visual observations. In the United Kingdom, observations of noctilucent clouds have been made by Paton (pers. commun. 1966) since 1939. In addition, observations have been made both by regular meteorological stations and by volunteer observers. In Germany, in the years after 1885, there existed under the direction of Jesse, Archenhold and Förster a working group in the then existing "*Vereinigung der Freunde der kosmischen Astronomie und kosmischen Physik*" (VAP) but for many years only sporadic observations were made. Vasilyev (1967) has published a study covering the early years of observations in the USSR (cf. Astapowitsch 1959; Bessonova 1963; Gromova 1963).

Attempts to find a correlation between the frequency of occurrence of noctilucent clouds and magnetic storms (type Sc) have been published by Willmann (1962). Bezrukova (1967) analyzed the relationship between the appearances of noctilucent clouds and sunspot activity. The results indicated that the changes in the Earth's atmosphere have a cyclic nature very similar to that of the 11-year cycle of sunspot activity. Vasilyev (1970) established a connection between noctilucent cloud displays and solar activity; a 4-year cycle of noctilucent cloud displays was discovered. He presented a tentative prediction of the number of nights during which noctilucent clouds can be observed (cf. Schröder 1970).

Using a complete list of the reported noctilucent clouds for the years 1885–1964, Fogle and Haurwitz (1966) have examined the suggested correlation between the clouds and solar activity. From their analysis no reliable conclusion as to whether there is a dependence of noctilucent clouds and solar activity is possible. As Fogle and Haurwitz have pointed out, an extended series of noctilucent cloud observations over a few sunspot cycles by the existing large network of observers and stations should provide the necessary data to settle this question.

D'Angelo and Ungstrup (1976) studied the occurrence of "widely observed" noctilucent clouds using the North American data of Fogle (1966), which encompass a sunspot minimum. Not one of these noctilucent cloud displays occurred on a day when the sum of K_p exceeded 15. D'Angelo and Ungstrup concluded that heating of the atmosphere produced by ionospheric electric fields and/or particle precipitation can be sufficient to reduce or to prevent formation of noctilucent clouds.

Gadsden (1985) found that this relation with the low Kp sums is not demonstrable in the NW European data. Gadsden showed that there were both long-term (based on sunspot numbers) and short-term (3-hourly Kp-index) connections between the frequency of occurrence of noctilucent clouds and solar-terrestrial activity. The relation is in the sense that when solar activity increases, there is a decrease in the frequency of occurrence of noctilucent clouds, which may be due to the atmosphere being warmed appreciably by solar activity.

The influence of solar activity on the lowest parts of the ionosphere (the D-region) is well established. For example, Lauter (1962) has analyzed measurements of radio absorption at low frequencies (245 kHz) for sunspot cycle effects. He found that the difference in absorption between sunspot maximum and sunspot minimum periods is a function of solar zenith angle χ. There is a pronounced maximum (a difference of 15 dB) at $\chi = 70°$.

7.3 Seasonal Frequency of Noctilucent Clouds

A necessary prerequisite for the treatment of noctilucent clouds is a comprehensive catalogue. Unfortunately, observations of noctilucent clouds have, up to now, been catalogued only sparsely so that very few compilations are currently available. The work of Soviet observers is compiled in several catalogues (Astapovich 1961; Bessonova 1963; Gromova 1963; Vasilyev 1967). The British observations are published in a series of annual reports (Paton 1964–1973; McIntosh and Hallissey 1974–1983; Gavine 1983–1987). Data from Germany are given in Schröder (1967c,e, 1968c); the North American observations are available in Fogle (1966a and Christie 1966, 1969a).

The degree of reliability of the observations is of decisive importance. For this it must be considered that the data are affected by local meteorological conditions and by the distribution of the observation stations. Although in the United Kingdom, the USSR and USA-Canada and Germany, routine observations are made, they are not an absolute guarantee of continuous monitoring; it is, therefore, sometimes impossible to make reliable statements as to the actual time at which a visual observation of the noctilucent clouds was made. And very often it cannot be seen from the reports on which (clear) nights *no* noctilucent clouds were seen.

Last, but not least, it is desirable to note the inhomogeneity of the observation network. In Germany at Roennebeck ($\phi = 53.2°N$) a regular watch has existed for aurora and noctilucent clouds since 1957 (Schröder 1962). In southern Europe no such network exists; only in the region of the USSR does a comprehensive network of stations exist so that some degree of continuity is guaranteed (Vasilyev 1967; cf. Astapowitsch 1959).

From his first analysis, Jesse established that the noctilucent clouds appear chiefly in the summer months in the northern hemisphere. Jesse and co-workers observed noctilucent clouds only during the period of May to August (Archenhold 1928).

When comparing the data obtained from the United Kingdom, the USA, Germany and the USSR, a difference can be seen. From early in August, no noctilucent clouds are recorded as visible from the United Kingdom (Paton 1964). For Germany, some are still seen as late as mid-August. Paton (pers. commun. 1966) remarked

Our observations of the occurrence of noctilucent clouds, using (data) from Britain, Scandinavia and Denmark show that the clouds are never seen before May 26th and seldom after August 3rd.

A recent analysis by Simmons and McIntosh (1983) has shown that the occurrence is limited to between mid-May and mid-August with median data in the first week of July (cf. also Fogle 1968; Schröder 1967d). Figure 7.1 shows histograms plotted from the data in three major lists of observations (Gadsden 1982).

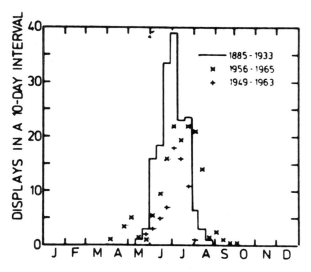

Fig. 7.1. The seasonal variation of noctilucent clouds in
the northern hemisphere. For
details of the data sets, see
Gadsden (1982)

From the USSR, there are repeated reports regarding observations during the months of October/November. In earlier German reports, we find repeated unequivocal reports of appearances of noctilucent clouds in the months of September even though their number is relatively small. There are occasional reports of sightings during the winter months but these should not be accepted as genuine sightings of noctilucent clouds (Schröder 1967c). It has been shown also that a winter observation by Hamilton (1964) was not an actual noctilucent cloud (Paton pers. commun. 1966; Schröder 1968b; Gadsden 1982). Hamilton reported for his data from Lerwick ($\phi = 60°N$) in January:

At about 16:55 UT the depression of the Sun was 10°45′. If it is assumed that the cloud was overhead and just illuminated at 16:55 and the refraction was twice 34′..., then the calculated height was 91 km — somewhat higher than the mean height of 82 km in the summer — and the cloud velocity was roughly 40 m s⁻¹ towards the south east.

It is of special importance for the theory of the noctilucent clouds to resolve the question of the appearance in winter. In Germany, in addition to noctilucent clouds, aurora borealis and the appearance of intensified night airglow are recorded. These programmes are carried out at Roennebeck ($\phi = 53.2°N$) during the entire year. Reference should also be made to the US observations and the statistics of intensified night airglow (Hoffmeister 1946; Schröder 1962) which cover a period of over 20 years. When we examine all these results, independent from each other, we can conclude that during winter months no noctilucent clouds appear on the northern hemisphere. The possibility that, for one reason or another, noctilucent clouds were "overlooked" may also be excluded. If noctilucent clouds did really occur in winter at any time at all one of these groups in Germany, United Kingdom or USA/Canada would have been sure to detect them (see Schröder 1966b, 1967b, 1967d, 1975).

It follows from these remarks that the noctilucent clouds are strictly a summer phenomenon. It has been assumed, therefore, that it is only during the summer that specific conditions prevail which result in the formation of noctilucent clouds (Schröder 1968a; Theon et al. 1966, 1967, 1969a,b).

The existence of noctilucent clouds in the southern hemisphere has been in doubt for some time because during the years 1885–1964 only three observers reported seeing noctilucent clouds in the southern hemisphere (Jesse 1889; Fogle 1965). From the northern hemisphere data, one would expect that noctilucent clouds occurring in southern latitudes would most likely be observed in the months December, January and February (with the peak of activity in January at latitudes of 50–60°S). In 1965, Fogle went to Punta Arenas, Chile ($\phi = 53.1°S; \lambda = 71°W$). During the days of January 8–17, seven nights were clear enough to see noctilucent clouds if they were present. He observed one display on the night of 9 January.

Additional observations were obtained during the summer 1965–1966 (Fogle 1966). On this excursion, noctilucent clouds were looked for during the period 27 November 1965 to 17 February 1966. Noctilucent clouds were seen on nine nights. They were observed to drift generally towards WNW. Figure 7.2 shows the percentage of clear nights with noctilucent clouds per ten days interval at Punta Arenas during the austral summer of 1965–1966.

Archenhold (pers. commun. 1985), has argued that there are *preferred dates* for the appearance of noctilucent clouds and pointed out that his father, in 1910, was the first to mention that comets (and their dust clouds) may be connected with the occurrence of noctilucent clouds. Briefly, Archenhold suggested that there are certain dates on which noctilucent clouds are more intense and visible over a wider area than usual. Such dates are, for example, close to

June 16: 1935; 1936; 1951; 1952; 1960; 1961; 1985
and
June 30: 1908; 1935; 1936; 1953; 1961; 1962; 1963; 1964; 1968; 1969; 1984.

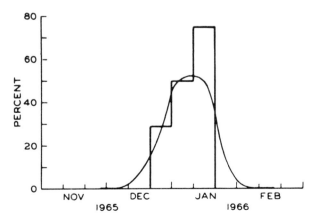

Fig. 7.2. Percent of clear nights with noctilucent clouds at Punta Arenas, Chile, during the austral summer of 1965–1966 (Fogle 1965)

This is a result of the Earth encountering dust particle activity on or near the path of a periodic comet.

The European data discussed above have been examined by Gadsden (unpublished) for the existence of such preferred dates. The "bright or extensive" displays were listed according to day of the year and the number of occurrences counted on a particular day. The noctilucent cloud season was taken to be 102 days long (from May 15 to August 24). There were 50 displays in total, so the expectation was 0.490. The results are given in Table 7.1 and show no significant departure from what is expected from the Poisson distribution. There appears to be *no* clustering around preferred days, or dates.

7.4 Climatology of the Mesosphere

It follows from long-term noctilucent cloud observations that they are strictly a summer phenomenon in the northern hemisphere. It has to be assumed, therefore, that it is only during the summer that specific conditions prevail which result in the formation and genesis of the noctilucent clouds (cf. Christie 1969a).

The physical behaviour of the mesosphere has been known only in the last few years. Haurwitz (1964), Ebel (1984), Hesstvedt (1964), Holton (1975), Houghton (1978), Kochanski (1963), Leovy (1964) and Memmesheimer (1985) have pointed out the inadequacy of knowledge in the field of mesospheric physics. In fact, this inadequate knowledge means that proof of various views cannot be carried out with the desired precision. In many cases one must depend on assumptions which may be quickly proven false by measurements.

In situ investigation of winds and temperature in the mesopause region at the time of a noctilucent cloud has been seldom carried out. Theon et al. (1969b) published results obtained by grenade sounding carried out from eight rocket launchings at high latitude sites in the summer of 1963–1965. The wind data were obtained at heights up to 95 km. These results show no obvious effect of a

Table 7.1. Clustering of "bright or extensive" noctilucent cloud displays around certain dates

Comparison of number of noctilucent cloud displays on a particular day of the year with the Poisson distribution.

Number of displays	Days on which this happened	Predicted from Poisson distribution
0	68	63
1	23	31
2	7	7.5
3	3	1.2
4	1	0.15
5	0	0.01
Total number of displays:	50	
Number of days in season:	102	
Expectation, Z, of Poisson distribution = 0.490		

cloud at or near the mesopause. They found an indication that noctilucent clouds are associated with lower wind speed than is the case when no noctilucent clouds appear (Table 7.2).

Figure 7.3 shows the course of temperature at the various height levels of the atmosphere. This average course of temperature shows a steep decline during summer in the region of mesopause. A few measurements of the temperature profile of noctilucent clouds were recorded during the 1962 ascents. A value of $T_{min} = 140$ K was obtained when noctilucent clouds were present and one of

Table 7.2. Observed mesopause winds in the presence and absence of noctilucent clouds

Site latitude	Date	Local time	Wind vector m s⁻¹ (degs)	Cloud
66°N	1963 Jul 30	01:28	31(101)	Moderate
66°N	1964 Aug 7	02:16	36(328)	Strong
66°N	1964 Aug 16	03:13	84(89)	Weak
66°N	1964 Aug 17	02:49	24(352)	Moderate
71°N	1965 Aug 7	01:13	29(267)	Strong
66°N	1963 Aug 2	01:27	96(53)	None
66°N	1963 Aug 8	00:29	165 (32)	None
71°N	1965 Aug 9	00:10	54 (72)	None

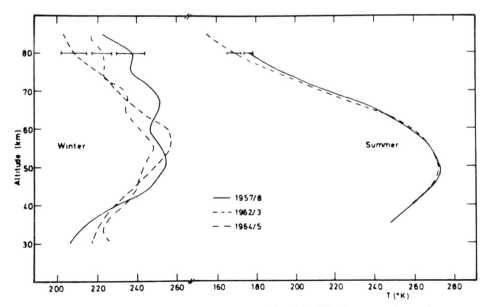

Fig. 7.3. Variation of mean temperature profiles at Fort Churchill (59°N) between solar maximum and solar minimum. For details see Smith et al. (1967) and Groves (1968)

$T_{min} = 140$ K when they were absent. Witt (1968) presented the results of six flights in 1963 and 1964, four were made when noctilucent clouds were present, two in their absence. The measurements extend from below 40 km altitude to 100 km and demonstrate that the temperature is not obviously different during a noctilucent cloud display from that without a display. This is shown also in the data of Theon et al. (1967).

As to the significance of the temperature value in the formation of noctilucent clouds, Theon et al. (1969b) stated:

Thus the coldest temperatures did not necessarily produce noctilucent clouds, but the clouds were always accompanied by mesopause temperature less than 150 K.

Witt (1968) presented the results of six flights in 1963 and 1964 (four in the presence, two in the absence of noctilucent clouds). His data showed that the atmospheric temperature is not obviously different in the presence or in the absence of noctilucent clouds.

There are two different atmospheric heat balance systems in winter and in summer. The middle atmosphere region (20–80 km) changes its dynamic structure rather drastically twice a year (cf. Ebel 1984; Lauter 1974; Gärtner, Memmesheimer and Blum 1986). There are significant reversals of the circumpolar zonal wind in spring and autumn. The spring transition starts from an existing state of rather strong coupling of the atmosphere layers from below by the propagation of planetary waves in the common wind systems (winds blowing from the west) of the stratosphere and the mesosphere. The spring reversal may start as a sudden breakdown of the polar vortex (Scherhag 1969) which rather abruptly terminates the upward propagation of planetary waves. The autumn transition arises from a slow and steady weakening of the stable, middle atmosphere, summertime, wind system (winds blowing from the east), mainly controlled by the heat balance of the upper stratosphere. The dynamical behaviour of the mesopause region is important for the occurrence of noctilucent clouds (Schröder 1968d, 1971).

Groves (1969, 1971) has shown that the zonally averaged meridional winds appear to be of the same order of magnitude as mean zonal winds in the lower thermosphere and upper mesosphere (see also Memmesheimer and Blum 1987).

A mass flux from the summer hemisphere to the winter hemisphere exists all the time at the level of the mesopause. Zonally averaged models have been widely used to study the large-scale circulation and related transport processes of the Earth's middle atmosphere. Memmesheimer and Blum (1987) presented a global, zonally averaged dynamical model of the middle atmosphere to calculate the seasonal variations of wind and temperature from 10–110 km. From their investigations it follows that changes around May 25 in the spring, and around July 4 in the autumn, are related to the presence of an established easterly wind jet in the lower mesosphere. Circulation changes in relation to the formation of noctilucent clouds are discussed by Schröder (1968d, 1971), who argued that there is a relationship between the change to summer circulation at lower altitudes and the onset of the noctilucent cloud season. This is seen in the model of Memmesheimer and Blum (1987) to some extent: the summer easterly jet in the mesosphere appears between May 10 and May 20 and vanishes between August 8 and August 18. There seems to be a time lag between the start of the noctilucent cloud season and the onset of easterly winds of about 10 days. On the other hand, the appearance of the noctilucent cloud season may be simply the last typical summer feature to appear in the mesosphere, arriving about 10 days after the easterly wind jet has been established lower down, and being the first summer feature to vanish, about 10 days before the breakdown of the summer easterly jet (cf. Memmesheimer and Blum 1988; Schröder 1968a,b).

Figure 7.4 uses the data of Theon et al. (1969a) to show the altitude of the boundary separating the disturbed region (upper curve) from the undisturbed region as a function of date. Theon et al. (1969a) found that the disturbed features typical of winter appeared in the upper mesosphere in September, and at progressively lower altitudes as time goes by (cf. Fig. 7.5). By December the full winter structure has been established in the northern hemisphere. In early January a warming at the stratopause begins a chain of events which eventually will restore the summertime structure.

Lauter and Entzian (1982) reported that during the changes from winter to summer, the wind system in the mesosphere disappears. At the same time, the ionization of the lower ionosphere attains a characteristic minimum. This D-region singularity can be used for defining the spring transition date at 80 to 86 km (see also Taubenheim 1975; Vincent and Ball 1984).

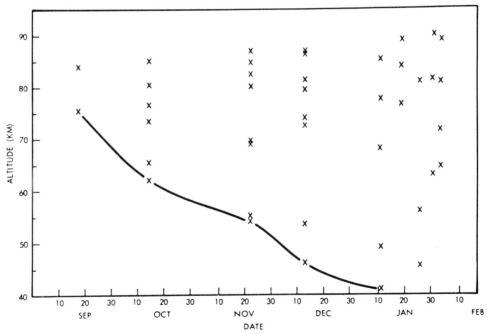

Fig. 7.4. The altitude of each lapse rate sign reversal (except the normal stratopause and mesopause reversals) during the transition period (Theon and Smith 1969a)

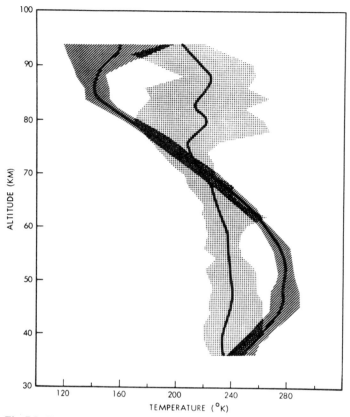

Fig. 7.5. Seasonal average temperature profiles above Barrow (Alaska 71°N). The *solid curve* is the average of ten summer soundings, and the *broken curve* the average of twelve winter soundings conducted during 1965–67. The *cross-hatched* areas surrounding each curve represent the total range of temperature included in the average (Theon et al. 1969a)

Apparently significant changes in all parts are observed during the transition to winter circulation (cf. Borbély 1972; Ebel 1974, 1984; Theon and Smith 1969). Similar variations are seen in the mesopause region, where noctilucent clouds appear in a very restricted height interval and noctilucent clouds can serve as an indicator of these changes. A picture emerges of common variations in the upper atmosphere in both spring and autumn (see Fig. 7.5). In this connection the seasonal and day to day changes are important for the diurnal variation of noctilucent clouds. A notable result appeared from the oldest data: the observers reported that the clouds appeared most frequently after midnight. Checking these older communications shows, however, that this is not the case. Apparently the terms "brightness" and "frequency" were exchanged. From the total German data from 1885–1956 it is clear that the time of most appearances of noctilucent clouds was before midnight (cf. Schröder 1966b, 1968c; see also Jensen and Thomas 1989).

8 Other Observations

8.1 Introduction

The region in which noctilucent clouds occur is part of the upper atmosphere which is chemically active and which is also open to much disturbance from outbreaks of solar activity. In determining the cause of noctilucent clouds, we can use both the regular emission of light (airglow) for diagnosis and the sporadic disturbances (aurorae) as tests of prognosis.

The airglow is emission of light during chemical reactions which take place in the upper atmosphere. The energy to create the reagents in the first place comes from sunlight. Radiation from the Sun, particularly in the ultraviolet region, breaks up atmospheric molecules during the day into single atoms and radicals. An example is provided by what happens to water vapour in the atmosphere. Above the stratosphere, solar radiation of wavelengths shorter than 239 nm can dissociate a water molecule into a hydrogen atom and a hydroxyl radical (OH). The hydrogen atom may then react with an ozone molecule (itself the result of a series of photochemical and chemical reactions) to give another hydroxyl radical and an oxygen molecule. This hydroxyl radical is in an excited state and can emit radiation in the visible and infrared parts of the spectrum.

The initial photodissociation of atmospheric molecules occurs during the daytime, of course, and ceases at night. The subsequent chemical reactions may take many hours or even days to occur; the pressure in the region is low and the reagents take time to come together. As a result, the emission of light goes on both by day and by night. From ground level, sunlight scattered from the troposphere and stratosphere tends to blind photometers used to observe the airglow and studies of the airglow are usually limited to nighttime or deep twilight.

The diagnostic value of airglow observations lies in having fairly well-known heights for particular emissions. Thus, the green line of atomic oxygen (557.7 nm) is emitted principally from a layer between about 90 and 100 km in altitude, with a peak at about 95 km. The hydroxyl emission (which consists of many bands from around 500 nm to beyond 2000 nm) comes from about 10 km lower, i.e. from 80–90 km with a peak close to 85 km. It is thus sometimes located at the top of the mesosphere (below the mesopause) but is more likely to be in the lowermost part of the thermosphere. There is an infrared emission (1.27 and 1.58 μm) of molecular oxygen from the 80–95 km region. The resonance doublet

(589.0 and 589.6 nm) of atomic sodium is present in the airglow and comes from close to 90 km; this particular radiation is very prominent in the spectrum of the middle twilight when the sodium atoms scatter sunlight resonantly.

Patchiness of the airglow can be used to look at movements of the upper atmosphere by simply watching, from ground level, the movements of any structure visible in the airglow (Clairemidi et al. 1985). The doppler width and shift of the atomic line at 557.7 nm are used to measure the temperature and the radial component of wind in the atmosphere at 90–100 km. The rotational structure of the emission bands of hydroxyl can be used to estimate the temperature of the atmosphere at 80–90 km. None of these measurements is good for measurement of exact temperature and wind components because they are for an indeterminate range of heights and represent an average value over the height interval. Nevertheless, the airglow measurements provide useful diagnostic tools for observing changes in the upper atmosphere from hour to hour and night to night. The intensities of the airglow emissions do not generally show dramatically rapid variations but their relative constancy does provide a check for accepting models proposed to explain the behaviour of the upper atmosphere.

With reference to Fig. 3.5, it is shown that there are regions (not geographically symmetrical) around each of the Earth's poles in which aurorae are seen quite frequently. Over North America and over Australasia, these regions overlap the latitudes from which noctilucent clouds are easily seen. Although aurorae occur continuously in two oval belts around the geomagnetic poles, they spread out to lower latitudes when the extreme outer atmosphere of the Sun (the solar wind) is disturbed. The effect in the noctilucent cloud belt is that there are sporadic bursts of keV electrons and protons into the upper atmosphere producing auroral displays. These particles come to a stop at about 105-km altitude although on occasion the lower border of an aurora may be as low as 80–85 km. At the base of an aurora, many electron-ion pairs are created through collision, of the electrons in particular, with atmospheric atoms and molecules. There is significant heating of the atmosphere. Data discussed by Banks (1977) suggest that the majority of the heating by particle impact is above 100 km in a typical aurora. At 90-km altitude, the energy seemed on one occasion to be almost as much as 10^{-7} eV s^{-1} for each and every atmospheric particle. This heating would produce a rate of temperature rise of no more than 4 K h^{-1}. Banks pointed out that auroral Joule heating (resistive dissipation of electric currents) in the upper atmosphere is as large as the heating caused by charged particle impact, but Joule heating is the more important heat source at 120–130 km altitude. It is negligible at altitudes below 95 km.

The appearance together of noctilucent clouds and an aurora has always caught the attention of observers. On July 24, 1950, Paton (1950) photographed from Abernethy (56°N) an auroral arc with its base at 10° elevation and noctilucent clouds reaching 5° elevation beneath the arc. The geographical positions of the two must have been very close. The clouds were watched and photographed from just after 2200 UT until they disappeared into the dawn at about 0309 UT. The display extended over the sky and at 0050 UT covered an area between azimuths 345° and 85°, up to an elevation angle of nearly 25°. Meanwhile, the aurora had vanished or had been overpowered by the brilliant light of the clouds.

In a later paper (Paton 1953), the aurora was described as developing sunlit rays in the early part of the night. At about 0205 UT, turbulence appeared in the eastern part of the cloud display and within a quarter of an hour, the clouds became

... quite chaotic in structure. . . though as the photograph shows, the parallel horizontal bands were still visible. Among some twenty displays of noctilucent clouds that have been observed since 1939, this is the only occasion when such turbulence has been seen. That it occurred soon after a brilliant aurora may be significant.

Did it arise from some thermal effect associated with aurora or was it merely the turbulence that would be expected near the earth-shadow boundary? At all events, it persisted unabated until the clouds vanished against the brightening sky at 0309 GMT. Sunrise was at 0402 GMT.

Whether an aurora occurring over a noctilucent cloud has a specific and definite effect on the cloud is still unclear. There is no agreement on what effect should be expected (see Schröder 1970).

8.2 Association with Hydroxyl Airglow Emission

Shefov (1967) has made a long series of measurements of the nightglow emission coming from excited hydroxyl radicals in the mesosphere. The altitude of the emission is quite close to that of noctilucent clouds and the radicals are excited in the process of their formation from atomic hydrogen and ozone. The atomic hydrogen available at those levels comes largely from the photodissociation of water vapour (see, for example, the discussion by Anderson and Donahue 1975). There are thus two principal reasons for directing attention to the hydroxyl emission in the context of observing noctilucent clouds. The rotational levels of the radical will be in thermodynamic equilibrium with the surrounding atmosphere and the rotational temperature should be the same as the local atmospheric temperature. The emission has a relatively open rotational structure and the determination of rotational temperature is therefore a relatively straightforward problem of spectrometry. Secondly, the intensity of the hydroxyl emission can be expected to be related to the water vapour content of the mesosphere, although the relationship need not be straightforward because the hydrogen:ozone reaction is only one of many involving both water vapour and ozone.

As one would expect, the rotational temperatures measured during the summer from Moscow (56°N) are low, in the vicinity of 160 K. When noctilucent clouds appeared, the temperature did not necessarily fall to a lower value; indeed, Shefov noted that sometimes the temperatures were actually higher than usual, with the lower temperature being seen the day after the display of clouds. The intensity of the hydroxyl emission was normally up by a factor of two during a cloud display compared with the average value on other nights during the summer. He saw evidence, too, that the night following a cloud display showed a decrease in intensity, to below the average by a factor of two or three, with a recovery to the average value by the second night after a display.

Harrison (1973), observing during one summer from Calgary (51°N) in Canada, obtained data from a single display of clouds together with data from other (cloud-free) nights in June and July. He could see no particular effect of the cloud upon either the temperature or the intensity of the hydroxyl emission.

Shefov's results are interesting but the interpretation is not necessarily straightforward. It has to be remembered that hydroxyl emission comes from a considerable thickness of atmosphere, probably extending at least 10 km in altitude. The rotational temperature is therefore a weighted average over a considerable range of temperatures. Furthermore, a change in measured temperature might as well be the result of a change in the mean height of the layer as an actual change in temperature of the mesosphere or lower thermosphere. Similarly, an increase in the intensity of the emission may well result from a decrease in the mean height of the emitting layer with consequent extra pressure at the airglow emitting layer. Perhaps, then, the changes of temperature and intensity associated with cloud displays come about through the changing height of the airglow layer. Whether the layer is higher or lower when the temperature is lower depends on the position of the mesopause relative to the airglow layer. The effects could come about through the distribution of water vapour in the mesosphere being drastically redistributed by water molecules freezing into ice particles during the night with subsequent daytime evaporation (in a narrow interval of height) to form a wet top to the mesosphere.

8.3 Association with Aurora and Planetary Magnetic Activity

Soon after noctilucent clouds had been recognized as something rather unusual, Smyth (1886), as we have seen earlier (sect. 4.2), noted the occurrence at Edinburgh (56°N) on one evening of both the clouds and an auroral arc. The same aurora was seen by Backhouse (1886b) from Sunderland (55°N); he added the information that the noctilucent clouds were visible chiefly before the aurora appeared and after it vanished; he reported that there is no reason to suppose there is any connection between the two phenomena. In a typed note found among J. Paton's papers (Byrne 1964), another simultaneous occurrence of aurora and noctilucent clouds is reported from Lerwick (60°N) in the Shetland Isles. On the night of July 30, 1963, the clouds were first noted at 2245 UT, towards west-north-west, covering about one-sixteenth of the sky. The elevation is given as 18° in the north-west, 26° in the west. At midnight, the clouds were barely discernible but by then there was an auroral arc visible towards the north-north-east, with a maximum elevation of 6°. At 0300 UT, the arc had rays and the noctilucent clouds became brighter. The cloud was in the same position as at 2245 UT, but had become more patchy and less fibrous than before.

Schröder (1965, 1970) saw no unusual changes in the clouds when aurora occurred on two nights in July 1963. The observations were made from Rönnebeck (53°N Germany).

Fogle (1966) arranged for a special observing programme in western Canada in the summer of 1965. Noctilucent clouds and aurora were observed together on 13 nights. On seven of the nights, they were in the same part of the sky. The general effect of the aurora was a decrease in the area and intensity of the clouds (on two occasions, they vanished) and a decay of their "well-ordered structure". Fogle suggested the possibility of auroral heating at or below the mesopause.

In this connection, however, it should be remarked that D'Angelo and Ungstrup (1976) have taken some of the North American data listed by Fogle (1966) and compared the dates of occurrence of noctilucent clouds that were observed over a wide area in 1963, 1964 and 1965 with the dates when the daily sum of the magnetic K_p-index (a quasi-logarithmic scale for allotting geomagnetic disturbance into ten classes) fell below 10, so-called geomagnetically quiet days. They found that almost all 21 cases of widely observed clouds fell at or near a minimum in the daily sum. D'Angelo and Ungstrup pointed out that the K_p-index is a measure of the strength of the electric field in and near the auroral zone at ionospheric heights and suggested that the marked anticorrelation between K_p-sum and the occurrence of noctilucent clouds arises from local heating of the atmosphere by electric currents. In turn, this leads to the air temperature rising above the frost point at a height where otherwise a cloud would form.

Annual lists of noctilucent cloud observations over western Europe have been published for many years from Edinburgh University. For a period of 6 years, Paton (1966, 1967, 1968, 1969, 1970, 1971) included in the lists dates when noctilucent clouds were almost certainly not present, i.e. when the skies were clear enough at a large enough number of observatories for him to be reasonably sure there were indeed no noctilucent clouds in the area. These data (covering the years 1967 to 1972) provide a good base for a repetition of D'Angelo and Ungstrup's analysis.

Two lists of dates have been made; list A contains 47 dates on which extensive or bright displays were reported. This list includes all displays shown as bright, brilliant, extensive, spectacular or widespread. It does not include displays that were weak, pale or "glimpsed through gaps in cloud".

List B includes those dates (144 in all) on which Paton notes "No NLC". Each list of dates is compared with lists of daily K_p-sum to give the numbers shown in Table 8.1. A curious relation emerges: there *is* an association between daily K_p-sum and the probability of occurrence of noctilucent clouds but the association is with the daily sum *following the cloud display*, not preceding it. By the time yet another day comes round, the association is gone.

Table 8.1. Association between the occurrence of bright or extensive noctilucent clouds and the daily K_p-sum[a]

K_p-sum:	0–10	10–20	20–30	30–40	> 40
Day ending at midnight ($X^2 = 0.3$):					
NLC:	11.5	27.5	6	2	0
	(11.2)	(26.6)	(6.6)	(1.8)	(0.8)
No NLC:	34	80.5	21	5.5	
	(34.3)	(81.4)	(20.4)	(5.7)	(2.2)
Day starting at midnight ($X^2 = 11.0$):					
NLC:	13.5	30.5	3	0	0
	(10.6)	(26.2)	(7.6)	(2.1)	(0.5)
No NLC:	29.5	76	28	8.5	2
	(32.4)	(80.3)	(23.4)	(6.4)	(1.5)
The following day ($\chi^2 = 2.5$):					
NLC:	13.5	27	6	0.5	0
	(11.7)	(25.1)	(8.0)	(1.5)	(0.8)
No NLC:	34	75	26.5	5.5	3
	(35.8)	(76.9)	(23.5)	(3.5)	(2.2)

[a] The number in parentheses below each number of occurrences is the number expected to occur if the "NLC" and the "No NLC" distributions are identical.

X^2 is calculated for three classes in each case (0–10, 10–20 and > 20), i.e. with 2 degrees of freedom. The X^2-test is used to test the null hypothesis, viz. that there is no difference between the distribution of K_p-sums in lists A and B. For any statistically significant association between noctilucent cloud frequency and K_p-sum, i.e. for the null hypothesis to fail, X^2 must to be greater than 9.

The association is weak: D'Angelo and Ungstrup found that the majority of extensive displays over North America fell on days when the K_p-sum was less than 10. For the West European data, this is true of less than one-third of the displays, a proportion not significantly different from chance.

The daily sum of the K_p-indices is rather a coarse index. Noctilucent clouds listed by Paton were usually seen slightly after Greenwich midnight and the K_p-index for 0003 hours UT is therefore essentially a simultaneous measure. The comparison is shown at the beginning of Table 8.2 and there is no significant association.

It might be thought that a more appropriate comparison is with the highest K_p-index in, e.g., the preceding 48 h. This is because of the quasi-logarithmic scale of K_p so that the geomagnetic disturbance on a day containing a couple of $K_p = 8$ periods is far greater than that on a day filled with $K_p = 2$ intervals. The second block of entries in Table 8.2 shows that this idea cannot be supported.

It is probably appropriate at this point to mention the correlation between mesospheric temperature and K_p found by Seshamani (1977). He took the published temperature soundings from Fort Churchill (59°N) made over the period 1956 to 1969. He found a positive correlation (r = 0.6; 98% significance level) between temperatures in the mesosphere, lagging 15 h, and K_p. This was for soundings in the daytime only; there was no significance to the small negative

Table 8.2. Association between probability of occurrence of noctilucent clouds and the 3-hourly K_p-index[a]

	0	1	2	3	4	5	6	7 or 8
K_p-index, 00–03h ($X^2 = 2.6$):								
NLC:	3	19	13	10	2	0	0	0
	(2.5)	(17.0)	(12.8)	(10.1)	(3.2)	(1.0)	(0.0)	(0.5)
No NLC:	7	50	39	31	11	4	0	2
	(7.5)	(52.0)	(39.2)	(30.9)	(9.8)	(3.0)	(0.0)	(1.5)
Highest K_p-index in previous two days ($X^2 = 4.0$):								
NLC:	0	2	6	24	8	4	2	1
	(0.0)	(1.0)	(6.4)	(18.9)	(10.6)	(6.4)	(2.0)	(1.7)
No NLC:	0	2	20	53	35	22	6	6
	(0.0)	(3.0)	(19.6)	(58.0)	(32.4)	(19.6)	(6.0)	(5.3)
K_p-index, 06–18 h on previous day ($X^2 = 4.2$):								
NLC:	18	71	50	34	8	5	2	0
	(15.0)	(72.8)	(53.9)	(27.8)	(10.8)	(3.9)	(2.0)	(1.7)
No NLC:	43	225	169	79	36	11	6	7
	(46.0)	(223.2)	(165.1)	(85.2)	(33.2)	(12.1)	(6.0)	(5.3)
K_p-index, 06–18 h following display ($X^2 = 25.5$):								
NLC:	22	78	60	23	5	0	0	0
	(15.5)	(69.9)	(50.9)	(33.2)	(11.1)	(4.7)	(1.7)	(1.0)
No NLC:	41	206	147	112	40	19	7	4
	(47.5)	(214.1)	(156.1)	(101.8)	(33.9)	(14.3)	(5.3)	(3.0)

[a] As in Table 8.1, the expected number of occurrences is given in parentheses beneath each observed number.

correlation coefficient (-0.2) for nighttime soundings. He found a slope of the temperature-K curve equal to

$$\Delta T / \Delta K = 9.1 \text{ K}$$

in the 81–90 km layer. If the correlation of temperature with magnetic activity occurring some hours before is a physically real effect, it is not clear why the correlation should not appear with nighttime soundings as much as with daytime ones.

The noctilucent cloud data show, in fact, an association of the K_p-index with a lag in the opposite direction. The last two blocks of entries in Table 8.2 show that the frequency of cloud occurrence is associated with the K_p-indices in the daytime following the night on which the cloud display was seen.

In summary, it is not clear what effect an aurora has upon clouds below it in the mesosphere. The examples mentioned above seem to refer to clouds and aurora at the same, or very similar geographical positions; one should remember that the aurora base is probably some 20 to 25 km higher in the atmosphere than the clouds. There seems to be no doubt about the statistical relationship between magnetic activity and the likelihood of seeing clouds, but the detailed behaviour for individual occurrences differs from case to case. One reason for the difference in behaviour between North America and western Europe may be caused by the more frequent occurrence of aurorae at the observers' latitudes in North America (see Fig. 3.5).

8.4 Lunar Effects

Kropotkina and Shefov (1975) investigated whether lunar tides in the upper atmosphere cause a temperature oscillation large enough to affect the probability of cloud occurrence. They took 1103 reported sightings in the months of June and July over the years from 1885 to the then present day. The data were sorted according to lunar time, smoothed with sliding blocks of 2 or 3 h and plotted as the departure of observed frequency from the average. They found a semi-diurnal component of 15%, and a diurnal component of 5%. The standard deviation is, according to the authors, not more than 5%. If one refers to the relation between rotational temperature of the hydroxyl airglow emission and lunar time (Shefov 1967), there is an expected change in temperature of 15–20 K. The association of cloud appearance with the time of lunar day is thus consistent with the airglow data.

The Edinburgh data for western Europe contain a similar lunar effect. The dates of 192 "extensive" displays (see above) in the 17 years from 1964–1981 have been sorted into daily intervals of the age of the moon (i.e. time after New Moon). The data are shown in the histogram given in Fig. 8.1. The thin line in the histogram has been fitted to the data by the method of least squares; the equation to the line is

$$N = 0.59 \cos(2\pi/29.5)(d-5.0)$$
$$+ 0.87 \cos(2\pi/14.75)(d-25.0) + 6.48,$$

where d is the number of (Earth) days that have elapsed since New Moon.

Analysis of the 30 differences between the numbers predicted by the equation and the numbers shown in the histogram indicates that the standard deviation of the amplitudes is 0.28. The amplitude (0.87) of the semi-diurnal wave is certainly significant (less than 1% probability of occurring by chance); the amplitude (0.59) of the diurnal wave is possibly significant (5% probability of chance). These results are remarkably akin to those of Kropotkina and Shefov and we can conclude that the lunar effect is well established.

8.5 Lidar Observations

The availability of pulsed lasers with high power in principle offers a convenient way of measuring the scattering cross-section of a noctilucent cloud. In practice, the measurements are difficult. A noctilucent cloud is optically thin and does not offer a dense target to return the light pulses. In addition, there is the possibility that the cloud has structure and the returns, therefore, may vary greatly in a few minutes or less. It is not, therefore, always possible to apply long integration times to improve the photon statistics of the observations.

The range to a noctilucent cloud overhead is less than the range to the layer of alkali metal atoms found at about 95-km altitude in the atmosphere. The optical thickness of a noctilucent cloud is comparable with that found for the less

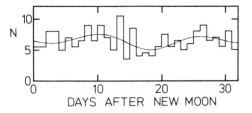

Fig. 8.1. Histogram of 192 occurrences of "extensive" noctilucent cloud displays seen over western Europe in 1964–1981 and reported in the Edinburgh lists. The *thin line* is a least-squares fit of a lunar diurnal (29.5 day period) and semi-diurnal (14.75 day period) oscillation

abundant alkali metals (potassium and lithium, but not sodium; Chanin 1984). French workers have solved the technical problems of detecting and measuring these less abundant alkali metals on a regular basis and have done so with great technical skill. Their observatory is, however, situated in southern France, well away from where noctilucent clouds occur overhead.

The only laser observations of noctilucent clouds appear to be those of Fiocco and Grams (1969). They used a laser emitting 100-ns pulses containing 2 J, with a maximum pulse repetition frequency of 0.5 s^{-1}. Observations were made from Kjeller (60°N) on 21 nights in July and August 1966. Noctilucent clouds were overhead on five evenings and present in the sky on a further four nights. Thus, there were 12 nights for comparison on which no noctilucent clouds were seen.

The measurements show that the 60–70 km altitude region contains an appreciable amount of particulate material. They found that the noctilucent cloud layer was at an altitude fairly close to 74 km. They reported changes in height of a particular cloud; in a period of 90 s, the height changed from 75 to 73 km. The geometrical thickness of the cloud appeared to be 0.5 km, with an optical thickness of approximately 10^{-4}. The photoelectron statistics of the data are poor, however, and the upper atmosphere scattering layers appear marginally above the returns from a dust-free atmosphere. It should be noted that the height given for this cloud is somewhat lower than that generally found by triangulation on other clouds.

8.6 Artificial Noctilucent Clouds

A bright "noctilucent cloud" was seen and photographed from Tucson (32°N) in the USA by Meinel et al. (1963). The height calculated from the time that the cloud vanished in twilight was 71 km. The figure is corrected for refraction but no mention is made of any screening height. The authors remarked that similar clouds have been seen on many occasions from Tucson; the days always coincided with the launch of a space vehicle from Vandenberg Air Force Base in southern California. They concluded that these clouds are the exhaust clouds of rockets.

Similar observations were reported by Benech and Dessens (1974). On two occasions, Feb. 23, 1971 and March 18, 1972, a mesospheric cloud was nucleated by the exhaust of a rocket passing through the region. The launch site was at Landes (44°N) in France; the cloud was formed slightly to the north of the site, close to 45°N on each occasion. Very little water was released by the rocket engines but a cloud which grew briefly was seen and the photographs of the cloud certainly show an appearance reminiscent of noctilucent clouds. The cloud on the second occasion was measured by triangulation and was found to be between 79 and 92 km altitude.

8.7 Abnormal Observations

Finally, mention should be made of observations of noctilucent clouds appearing under unusual circumstances, in odd places, or at the "wrong" time of the year.

First, there are observations of noctilucent clouds seen during a total eclipse of the Sun. Fast and Fast (1981) were in the small village of Krasnoyarka (55°N) to view the total solar eclipse of July 31, 1981. They saw noctilucent clouds on the evening before the eclipse and again for some 6 min during and after totality. The clouds were horizontal streaks extending about 30° to 40° in azimuth and no higher than 5° or 6° in elevation. They were seen in the azimuth of the eclipsed Sun (approximately 35° elevation), below the corona. The total phase had a duration of 84 s and the clouds, which were first noticed after totality had started, were visible for 4 to 5 min after third contact. They reported that the noctilucent clouds were confidently identified by more than ten observers who had experience with noctilucent cloud observations. The scattering angle for the observation lay between 30° and 40°, rather similar to that seen during normal twilight observations.

The second instance is that which appeared to be noctilucent clouds at *tropical* latitudes. Avakyan et al. (1981) discussed Salyut-6 observations which showed a scattering layer at 80–95 km altitude above the equator. This layer was seen by four astronauts in the period June 25 to July 5, 1979; they had seen noctilucent clouds at high latitudes on the same orbits. The equatorial layer was estimated, in a preliminary way, to have a vertical optical thickness of 10^{-4} to 10^{-3}; assuming particles of radius 0.3 μm, there were from 0.1 to 1 m^{-3}.

In a recent paper on these tropical "noctilucent clouds", Avakyan et al. (1986) presented English translations of on-board journals kept by Salyut-6 cosmonauts Kovalyonok and Savinykh, with descriptions of clouds seen in 1981. For instance, the cosmonauts saw (on March 22, 1981) a

... field of NLC above the western part of the Atlantic Ocean in the latitudes approximately 15°–35°N.

There appeared at this time to be a vast area of mesospheric cloud stretching from 75°W to 37°W between 15°N and 47°N. There were more observations reported in April and May of that year; on May 14, 1981, mesospheric clouds were seen over south-eastern Australia, during late autumn in the southern hemisphere.

An extreme example of noctilucent clouds being seen at the "wrong" time of the year has recently been reported. Griffiths (1987) photographed what are undoubtedly noctilucent clouds at Argentine Island, in the southern hemisphere, in *June*, that is, in the middle of the winter. The time of his observation was when the Sun was some 11° below the horizon, and the clouds were photographed in the direction of the solar meridian. The observations were so unexpected that Griffiths asked experienced observers to examine his colour pictures to verify that, indeed, the clouds had all the appearance and characteristics of noctilucent clouds. There must have been very unusual things happening at the top of the mesosphere in June 1985 over the South Atlantic (cf. Schröder 1988b).

9 Environment of Noctilucent Clouds

9.1 Introduction

The region of the atmosphere in which noctilucent clouds are formed is one of very low pressure in comparison with the pressures at which other types of clouds are formed. The physical processes of importance in discussing noctilucent clouds are correspondingly changed in importance. First, the mean free path in the region is much larger than the particle size in the cloud and the physical conditions are those of the Knudsen regime rather than those of a fluid. It may help in visualizing the local conditions to list some of the relevant properties of the atmosphere. Table 9.1 lists the mean free path of air molecules, their mean velocity, the number density and the flux passing a plane surface. In calculating the values given in Table 9.1, the cross-sectional area of a molecule was taken as 4.417×10^{-19} m^2, in accordance with viscosity data, and the mass of an air "molecule" was set equal to 4.845×10^{-26} kg.

9.2 Atmospheric Temperature

It is a fact that the temperature around the mesopause seems to be the same when noctilucent clouds are present and when they are not. This is shown very clearly in the rocket soundings made at Kronogard (Sweden) and at Barrow (Alaska) during the summers of 1963, 1964 and 1965. Witt (1968) presented the results of

Table 9.1. Kinetic properties of the atmosphere: model for illustrative purposes

Altitude	Temperature	Pressure	Mean free path	Mean molecular speed	Molecular number density	Molecular flux
z	T	p	λ	c	N	n
(km)	(K)	(Nm^{-2})	(mm)	(m s^{-1})	(m^{-3})	(m^{-2} s^{-1})
70	220	7.67	0.83	400	2.53×10^{21}	2.53×10^{23}
75	195	3.44	1.65	376	1.28×10^{21}	1.20×10^{23}
80	165	1.36	3.5	346	5.98×10^{20}	5.17×10^{22}
85	140	0.456	8.9	319	2.36×10^{20}	1.88×10^{22}
90	152	0.143	31	332	6.82×10^{19}	5.66×10^{21}

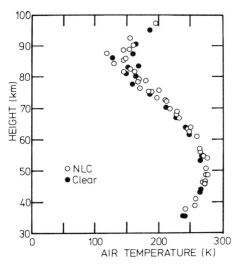

Fig. 9.1. Results of rocket soundings of atmospheric temperature made from the Kronogard (Sweden) rocket range. The *filled circles* are for times when there were no noctilucent clouds; the *open circles* for occasions when noctilucent clouds were seen in the vicinity of the rocket path

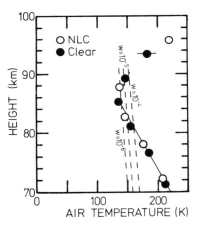

Fig. 9.2. Rocket-borne measurements of the mesopause over the Barrow (Alaska) range. The *oblique (broken) lines* show the variation with height of the frost point for water vapour mixing ratios equal to 10^{-4}, 10^{-5} and 10^{-6}. Here, and in Fig. 9.1, it is difficult to see any great difference in the temperature regime corresponding to either the presence or the absence of noctilucent clouds

six flights in 1963 and 1964, four of which were made with noctilucent clouds present, two in their absence. The data, redrawn here as Fig. 9.1, extend from below 40 km altitude to nearly 100 km and show clearly that the atmospheric temperature is not obviously different during a cloud display from what it is without a display. Similar results are shown in the Barrow data (Smith et al. 1967) which are plotted in Fig. 9.2. These latter data form the basis for the model temperatures used in Table 9.1. The model pressure used is based on a reference pressure at 80 km taken equal to that given in the CIRA 1965 reference atmosphere (COSPAR 1965) for July 1 at 60°N latitude.

For a review of the physical processes occurring in the mesosphere during summertime, the reader is referred to Björn (1984). He discusses at some length the results of rocket-borne soundings of temperature and their implications (cf. Schröder 1971; Theon et al. 1970).

9.3 D-Region

The region in which the clouds form is part of the ionospheric D-region. The atmosphere is partly ionized and contains large numbers of free electrons, positive (and some negative) ions and, of particular interest, ions on which numbers of water molecules have clustered. Reid (1977) has discussed in some detail how these different species of charged particles interact. He found that the electron density at 80–85 km altitude can be expected to be about 1.0×10^9 m^{-3} under normal, daytime conditions.

Normal, in this context, means there is no flux of high energy electrons being precipitated from the magnetosphere. Much of the high mid-latitude area at which noctilucent clouds occur is also one of high geomagnetic latitude and as such is open to fluxes of high energy particles.

Arnold and Joos (1979) reported the results of mass spectrometer measurements on three rocket ascents launched from Andoya (in northern Norway) and Kiruna (in northern Sweden). On two of the flights, with enhanced electron fluxes, the electron density at 82 km was found to be 2.0×10^{10} and 6.6×10^9 m^{-3} respectively, compared with 3.0×10^9 m^{-3} on the third flight when the electron fluxes were close to their background level for daytime.

The ionic chemistry of the region (Hunt 1971) predicts that the principal positive ions in the region will be hydrated NO + ions (the positive ion of nitric oxide with one to seven or more water molecules adhering to the initial ion). Reid (1977) showed that the relative amounts of the different hydrated ions will depend on both the temperature and the humidity of the atmosphere, as would be expected (see also L. Thomas 1976). Arnold and Krankowsky (1977) have used measurements of the hydrated ion to estimate that the water vapour mixing ratio in this region is close to 4.0×10^{-6} (by volume). The flights from which their data were taken were made at latitudes of 68° and 69°N. The rockets were launched in May and August and the measured temperature at 82 km was 170 K on the May flight and 166 K in August.

Goldberg and Witt (1977) flew a rocket-borne mass spectrometer on a day when noctilucent clouds had been seen over the launch site the previous night. Their data showed hydrated ions and they drew attention to the presence of ions which were possibly water clusters surrounding Fe and FeO initial ions. Whether this finding is peculiar to the atmosphere where noctilucent clouds have formed or whether it is a coincidence with enhanced meteor influx must be elucidated after more flights under these special conditions have been made.

Johannessen and Thrane (1974) reported the results of the flight of a mass spectrometer through the mesosphere over northern Norway in the month of

August. Up to 84 km, water clustered onto a proton dominated the ion composition. There was an abrupt transition in the region of 84–86 km, and the dominant ions above that region were found to be $NO+$ and $O+$ with negligible water clustering. The results are of particular interest because the rocket payload included ionization chambers which were used to estimate the temperature of the atmosphere from the extinction of solar Lyman-alpha. The temperature showed, on this occasion, two minima (a double mesopause) at 80 km (145 K) and at 85 or 86 km (150–155 K).

Yet more information concerning noctilucent clouds and hydrated ions comes from the data discussed by Björn et al. (1985) and Kopp et al. (1985b). Rocket-borne mass spectrometers showed a narrow layer of hydrated ions lying some 7 km above a "weak" noctilucent cloud which was detected at 83-km height by photometric observation. In interpreting such data, one should bear in mind always that there may be two or more distinct air masses present in the upper atmosphere and that one may be sliding over another. There is thus the possibility of decoupling in the vertical; what is going on at one level may be not particularly relevant to phenomena occurring one to two scale heights below.

As part of the "Cold Arctic Mesopause Project" in 1982 (Grossmann et al. 1985), a salvo of seven rockets was launched from Esrange (Sweden) at a time when noctilucent clouds were to be seen over the range. A second salvo of three rockets was launched 8 days later when observers reported the sky over the range to be clear of noctilucent clouds.

The first salvo was at a time of high magnetic activity in addition to the noctilucent cloud presence, and positive ion densities were measured to be 2.5×10^{11} m^{-3} at 90 km and 1.6×10^{10} m^{-3} at 80 km. The noctilucent cloud was measured to be at 83 km, where the temperature showed a minimum of 138 K. There were other minima seen on the measured temperature-height curve, one of 114 K at 89 km and another of 111 K at 94 km (Philbrick et al. 1984; Kopp et al. 1985a). On-board sensors of atomic oxygen suggested that there was a small depletion of atomic oxygen in the neighbourhood of the noctilucent cloud and it has been suggested that this is the result of increased association of atomic oxygen caused by water molecules evaporating from noctilucent cloud particles at the bottom of the cloud layer. (These oxygen observations are difficult to interpret. The measurements refer to the underside of the principal layer of atomic oxygen which peaks at approximately 95 km. There are minima in the oxygen density at 89 and 94 km also, corresponding to the other minima in temperature but not, presumably, also associated with increased water molecule concentrations.)

Ganguly (1984) has used VHF incoherent scatter spectra to observe from ground level the height at which ions become hydrated, that is, the transition from simple positive ions to water-cluster ions. The transition coincides with a ledge in electron density, a sudden increase in electron density with increasing height. There is a clear diurnal variation in this height which drops some 10 km towards noon, and rises again during the afternoon.

9.4 Dust

In a recent paper, Hunten et al. (1980) set up a model of the stratosphere and mesosphere in which they calculated the amount of dust and smoke that can be present from the ablation of meteors. Using a distribution of meteor velocities that has a mean of 14.5 km s^{-1}, they calculated that the majority of the meteoric material is deposited between heights of 75 and 95 km. They pointed out that the ablated material is unlikely to be in the form solely of boiled-off molecules; there will be "smoke" particles in the meteoric wake. These particles act as nuclei for the aggregation of some of the meteoric material and form a significant source of solid particles, presumably of siliceous material, at the top of the mesosphere. This is, of course, the region that contains the whole phenomenon of noctilucent clouds and they suggested that these solid particles may well be an important source of nuclei for the clouds. It is clear that the solid material itself cannot account for the clouds; the scattering cross-section of the material is far too small to account for the amount of sunlight scattered in a noctilucent cloud and, indeed, Hunten et al. estimated that the amount of sunlight scattered by the material is *undetectable*.

The numbers of nuclei that are present, according to the calculations, are critically dependent on the assumption of initial particle size in the meteor smoke. It is assumed that a mass influx of 1.0×10^{-15} kg m^{-2} s^{-1} is distributed among identical smoke particles of radius r_0 and density 2000 kg m^{-3}. Clearly, then, the number of initial nuclei is inversely proportional to r_0^3 and the steady-state dust particle concentration in the source region will vary quickly as the chosen value of r_0 is changed. For $r_0 = 0.2$ nm, there will be a dust concentration at 85 km of greater than 5.0×10^{10} m^{-3}; when r_0 is set five times larger (1.0 nm), the dust concentration falls to 3.0×10^9 m^{-3}. For the rather large initial radius of 10 nm, the dust concentration is down to approximately 1.0×10^6 m^{-3}. Coagulation is probably not very important in the 80–90 km region; by the time the dust has settled to 60 km, however, there are particles of radii up to several nanometres regardless of how small the value of r_0 has been assumed initially.

Solomon et al. (1982) have shown that meteoric ions are involved inextricably with the chemistry of water molecules. Indeed, determination of the number density of silicon ions in the lower thermosphere may, in principal, be used to estimate the local water vapour mixing ratio.

9.5 Water Vapour in the Mesosphere

Liu and Donahue (1974) have discussed in a comprehensive manner the processes that set the water vapour mixing ratio at any particular height in the upper atmosphere. To give a simple summary of their findings, there is a continuing upward flux of hydrogenic material. This flux is explicitly involved in the "boiling-off" or escape of hydrogen atoms from the outermost parts of the Earth's atmosphere.

There is a wet, lower boundary layer to the system, the troposphere. Water vapour diffuses upwards from this source and suffers photodissociation in the upper stratosphere and mesosphere. Nicolet (1984) gives a general discussion of this. Other hydrogenic material from the surface, methane for example, is also involved in this upward diffusion leading to photodissociation.

In the mesosphere, where there are atomic oxygen and ozone present, a complex system of chemical reactions (T.F. Thomas et al. 1979) controls the amount of water vapour at any particular place and time. Eventually, at thermospheric levels, all the atmospheric hydrogen is present either as atomic hydrogen or protons.

The temperature of the thermosphere, at heights of the order of 500 km is such that a significant number of the hydrogen atoms have thermal speeds great enough that, if moving in an upward direction, the atom will leave the Earth altogether on a hyperbolic path.

The region in which this occurs is called the *exosphere* and the level at which an atom has a probability equal to $1/e$ (0.37) of leaving the Earth is known as the *exobase*. For typical thermospheric conditions, the exobase for hydrogen is at approximately 460 km.

The exosphere acts as a sink for hydrogen in the atmosphere. To maintain the flow by diffusion of hydrogenic material going from the source (the troposphere) to the sink (the exosphere), it is necessary that the mixing ratio of the hydrogenic material shows a negative gradient with height. Liu and Donahue pointed out that the eddy diffusion coefficient in the stratosphere and mesosphere is so great that the escape flux is maintained with a negligibly small gradient of mixing ratio. The water vapour mixing ratio in the neighbourhood of noctilucent clouds is set, therefore, by chemical and ionic effects, together with bulk movement (meteorological processes).

Estimates of the amount of water vapour present in the vicinity of the mesopause usually put it at a few parts per million (Anderson and Donahue 1975; Arnold and Krankowsky 1977). Rather larger amounts have been given by the ground-based microwave measurements of Radford et al. (1977); the altitude distribution of atmospheric water vapour is obtained from mathematical inversion of the line profile of an atmospheric emission line (Deguchi and Muhleman 1982). Because of pressure broadening and Doppler (thermal) width, the contribution to the centre of the line profile seen at ground level is mainly from water vapour at high altitudes, while the water vapour lower in the atmosphere contributes to the entire profile. Mixing ratios are found to be as high as 15 ppm by volume in the 60–80 km region; the data do not give estimates any higher than 80 km. Recent work reported by Olivero (Olivero et al. 1981) gives 10 ppm at 65 km as typical of the atmosphere over the eastern USA (that is, at mid-latitudes).

The microwave measurements of Bevilacqua et al. (1985) and Olivero et al. (1986) suggest that there is a twofold increase in the water vapour mixing ratio at 75 km during the summer (measurements were made from April to June).

Hartmann et al. (1983) have overcome the problem of blanketing by tropospheric water vapour by flying a microwave receiver in an aircraft at or near the tropopause. The decrease in background level surrounding the spectral line permits usable data to be gathered in a shorter time interval, minutes rather than hours.

Fiocco and Grams (1971) discussed the possible effect of extraterrestrial dust (particles of minute radius that are stopped high in the atmosphere without ablating) sifting down into the mesopause region and sweeping water from the lower thermosphere down to just below the mesopause. They noted that the meridional trajectories of dust entering the atmosphere are likely to give a dust layer at roughly 70 km at high latitudes. Indeed, laser soundings have shown just this finding.

Fiocco and Grams (1971) suggested that the dust particles provide efficient nucleation of water vapour at a region above the mesopause and that the dust particles, with adsorbed water, simply scavenge water vapour in the upper levels down to below the mesopause, where the temperature is high enough to evaporate the water or ice from the dust. There may well be, therefore, an unusual amount of water vapour in the mesosphere at high latitudes which will not be seen in measurements made at mid-latitudes.

As we have seen in Section 8.5, laser soundings from southern Norway (Fiocco and Grams 1969) do indeed show increased scattering from the high atmosphere near 65 km and at 74 km. The association with the appearance of an overhead noctilucent cloud is not good, however. In spite of the 74-km layer appearing on one night concurrently with a noctilucent cloud, that layer shows up in the summer's observational summary (21 nights of observations: no noctilucent cloud on 12 nights, noctilucent cloud overhead on 5 nights and present elsewhere in the sky on a further 4 nights) as statistically above the signal fluctuations on only two nights, each without any noctilucent cloud display.

9.6 Radiation

Deirmendjian and Vestine (1959) considered the absorption and emission of radiation by cloud particles in sunlight. Using a very simple model, they found that the equilibrium temperature of an ice particle is greater than 273 K. This conclusion was criticized by Bronshten (1970) who considered the radiative equilibrium of a submicron particle which is irradiated both by the Sun and by thermal radiation from the Earth and its atmosphere. Cooling of the particle takes place by radiation and conduction through the ambient atmosphere. He found that the equilibrium temperature of an ice particle at 80-km altitude is 170 K, taking the local atmospheric temperature to be 160 K. This applies only if a particle has a radius no greater than 1 μm.

Fiocco et al. (1975) have tackled the radiative transfer calculations independently, using an aerosol model involving refractive indices larger than is the case for ice or water. In the visible, the refractive index is taken to be 1.65 and has an appreciable imaginary part in the near-infrared (i.e. there is infrared absorption). At 45°N, summertime, they found an excess of particle temperature above ambient (atmospheric) temperature equal to 175 K for 0.5 μm particles at 92 km. With the particle radius down to 0.01 μm, the excess reduces to between 1 and 10 K in the 80–90 km height interval. Such temperature excesses are not applicable to the case of ice particles.

Baibulatov and Ivaniya (1976) showed that the temperature excess at 80 km is less than 10 K for all particles of radii up to several micrometres. The difference between these numbers and those of Fiocco et al. comes from taking ice as being much more of a dielectric than the rather absorbing, high refractive, index material that Fiocco et al. considered. In a second paper, Baibulatov and Ivaniya (1977) considered in more detail the region of occurrence of noctilucent clouds and the temperatures of particles at these heights. They calculated the temperature excess for both pure ice particles and layered particles, i.e. ice with a core of absorbing material. The temperature of small particles (radii up to 0.1 μm) is essentially identical to the air temperature up to altitudes of close to 85 km. In this region, the temperatures of larger particles are higher in the summer and lower in the winter than the air temperature. In the case of the particles with a core, the effect of the core is marked if it occupies more than about one-seventh of the radius of the particle. The particles with a large central core have to be much smaller than 0.1 μm radius if they are not to have a large temperature excess.

Olivero and Bevilacqua (1979) considered the radiation balance established at the surface of ice particles. The particles absorb solar radiation (direct and reflected) and terrestrial radiation from below. Molecular thermal conduction and latent heat of sublimation are also taken into account. Simultaneous equations have to be solved for the rate of change of both particle temperature and particle radius; the authors found that quasi-equilibrium is reached within a few seconds. The steady-state temperatures are quite different in general from the temperature of the surroundings. For high latitude summer conditions, this difference appears to be small ($<$ 10 K) at the top of the mesosphere. If one defines an "existence region" as one in which the particles are not subliming, the extent in latitude of the existence region depends markedly upon the water vapour mixing ratio. For the existence region to extend from the pole to 67° latitude, a mixing ratio of 100 ppm is needed. Clearly, this rather high value of the mixing ratio can be reduced to a more acceptable value if the mesopause temperature is taken as lower than the 140 K assumed for these calculations.

Willmann et al. (1981) discussed the results of the Salyut-6 observations in the southern hemisphere in conjunction with simultaneous rocket soundings for temperature. The data refer to the period of December 1977 to February 1978. Noctilucent clouds were seen each day between 23 December and 2 February. There were resistance wire measurements (Izakov et al. 1967) of atmospheric temperature and wind velocities obtained from tracking of "window" ("chaff") at altitudes from 78 to 83 km. Some 14 rocket flights were made from a launch site at 68°S in the period from 7 December to 7 March. It is clear that the rocket soundings did not penetrate to sufficiently high levels to allow the mesopause to be properly defined. On the majority of the flights, the minimum in temperature was not traversed. The authors gave a summary of the temperature data for the height interval 78–84 km. The average (for 68°S) shows the temperature falling from 156 K at 78 km to 129 K at 84 km, with a lapse rate increasing over the same range from 7.0° to 26.5° km⁻¹.

9.7 Rates of Growth

For our purposes, probably the most interesting data are those relating to the speed of growth and sublimation of an ice crystal. The shape is probably not spherical; the departure from sphericity can be allowed for by using the concept of an inscribed sphere, which is given the radius R. If the surface area and volume of the crystal are A and V, two shape factors may be defined as

$$a = A/4\pi R^2 \tag{9.1}$$

and

$$b = 3V/4\pi R^3. \tag{9.2}$$

If the mass flux of water molecules per unit area at the surface of the crystal is given by g(T) at temperature T, then the rate of increase in mass of the crystal is clearly

$$dM/dt = 4\pi R^2 a \{g(T_c) - g(T_a)\} m, \tag{9.3}$$

in which T_a is the temperature of the atmosphere and T_c is the temperature of the crystal. The mass of a single water molecule is denoted by m. The value of $g(T_a)$ is given simply by the product of the water vapour mixing ratio, w, with the value of molecular flux, n, in Table 9.1. The value of $g(T_c)$, the flux of water molecules evaporating from the crystal at temperature T_c is found, for any T_c, by calculating n for the saturation vapour pressure of water at temperature T_c.

The saturation vapour pressure over ice at these exceedingly low temperatures must be estimated; it is too low to be measured with precision in the laboratory. Thus, use is made of the Clausius-Clapeyron relation

$$L_T = RT^2 d(\ln p)/dT, \tag{9.4}$$

in which L_T is latent heat of vaporization of ice at temperature T; p is the vapour pressure over ice and R is the universal gas constant. If C_s, C_g are the specific heats at constant pressure of ice and water vapour respectively, then

$$L_T = L_o + \int_o^T (C_g - C_s)dT. \tag{9.5}$$

The specific heat of water vapour, C_g, at low temperatures is listed by Eisenberg and Kauzmann (1969) and we can use the Giauque and Stout (1936) values for C_s, with 616 Nm⁻² for the value of p at 273 K. Numerical integration of the Clausius-Clapeyron equation in steps of 10 K gives values of p which are closely approximated by the following simplified version of the Kirchhoff formula:

$$\ln p = 28.548 - 6077.4/T, \tag{9.6}$$

in which p is in Nm^{-2}. If we now write ρ (912 kg m^{-3}) as the density of ice, the rate of change of the radius of the inscribed sphere can be written

$$dR/dt = (aM/b\,\rho)\,\{g(T_c) - g(T_a)\}. \tag{9.7}$$

There are two limiting cases for the rate of change of radius with time: first, the maximum rate of evaporation is when $g(T_a) = 0$, that is, when the cloud particle falls into a perfectly dry region of the atmosphere. The values of dR/dt in this case are independent of the atmospheric model and are related to the temperature of the particle only. Some numerical values are given in Table 9.2.

Similarly, there is a maximum rate of growth that occurs when $g(T_c) = 0$, that is, when the temperature of the cloud particle is sufficiently low that the rate of evaporation of water molecules from its surface is negligibly small. In this case, it is necessary to take account of the humidity of the atmosphere.

The water vapour mixing ratio in the upper atmosphere during noctilucent cloud displays is not known (see above), but is thought to be in the range, e.g. of 2.0×10^{-6} to 2.0×10^{-5}. With these values and the model atmosphere given in Table 9.1, it is possible to calculate the rate of change of the radius of the inscribed circle; these are given in Table 9.3. (Only the atmospheric conditions at 85 km are considered; at the other levels listed in Table 9.1, evaporation rather than growth takes place except with rather high values of the water vapour mixing ratio.) The third column in Table 9.3 shows the time taken for a spherical cloud particle to grow to a radius of $0.1 \mu m$.

Unless there is much more water vapour at the top of the mesosphere than there is currently thought to be, the growth of a noctilucent cloud particle is exceedingly slow. This rules out interpreting wave patterns in noctilucent clouds as being the consequence of temperature variations which cause evaporation (thinning) of the noctilucent cloud when the temperature rises, followed by re-assembly of the water molecules during the falling temperature part of a cycle. As we have seen, a few tens of degrees rise in temperature can certainly cause significant evaporation in a relatively short time but there cannot be an equally fast rebuilding of the noctilucent cloud particles.

Table 9.2. Maximum possible rate of sublimation and time taken for the disappearance of a sphere of initial radius equal to 0.1 μm

T (K)	(dR/dt) (m s^{-1})	Time
130	2.23×10^{-14}	52 days
150	1.05×10^{-11}	2.6 h
170	1.17×10^{-9}	85 s
190	4.74×10^{-8}	2 s

Table 9.3. Values for the rate of change of radius of a spherical crystal for particular values of the water vapour mixing ratio, w

T = 140 K; p = 0.456 Nm^{-2}		
w	(dR/dt) (m s^{-1})	Time (h)
2.0×10^{-6}	1.23×10^{-12}	23
5.0×10^{-6}	3.08×10^{-12}	9
1.0×10^{-5}	6.16×10^{-12}	4.5
2.0×10^{-5}	1.23×10^{-11}	2.3

9.8 Nucleation of Ice

Two things are involved in having a formation rate of embryonic particles that is large enough to account for the presence of a visible cloud. The partial pressure of water vapour must be high enough to reduce a free-energy barrier to an acceptably small level. There must be sufficient nuclei present to maintain a source of embryos against their loss by growth and consequent falling out of the saturated region. The degree of saturation, S, at any height in the atmosphere is simply the ratio of the local partial pressure of water vapour to the saturated vapour pressure over ice at the temperature, T, at that height. As a cloud particle grows, there is a change in the free energy, G, of the particle-vapour system. In the absence of any electrical charge on the embryo, there are two terms in the expression for the rate of change of G with change in radius (cf. Hobbs 1974):

$$dG/dr = 8\pi a\sigma r - 4\pi b\rho r^2 (RT/M) \log_e S, \tag{9.8}$$

in which M is the molecular weight of water and σ is the ice-vapour surface energy. If there should be an extra proton on the embryo, the Coulomb forces on an approaching water molecule cause an extra term to appear in the equation:

$$dG'/dr = dG/dr - \alpha \rho e^2/(8\pi\varepsilon_0^2 mr^2), \tag{9.9}$$

in which α [1.606×10^{-40} Cm/(Vm^{-1})] is the polarizability of a water molecule; ρ (932 kg m^{-3} at 130 K) is the density of ice; e = 1.602×10^{-19} C; ε_0 = 8.854×10^{-12} Fm^{-1}; and m = 2.99×10^{-26} kg, the mass of a water molecule.

The first equation may be integrated immediately; dG'/dr may be integrated with the value of G' set to that of G for very large particles. The resulting curves of G (light lines) and G' (bold lines) are plotted in Fig. 9.3. A temperature of 130 K is assumed and 0.109 J m^{-2} is used for the surface energy.

In obtaining the expression for G', no account is taken of a water molecule changing its orientation as it approaches the charged embryo; there is then no need to take account of the permanent dipole moment of water in the expression for dG'/dr.

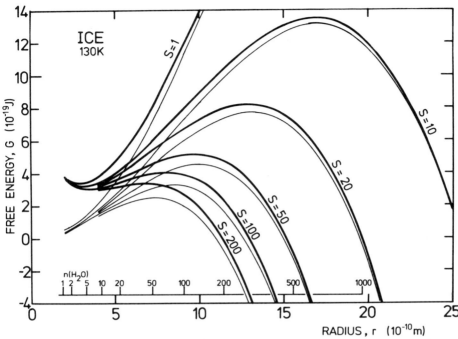

Fig. 9.3. Calculated free-energy barriers governing the growth of an ice crystal for various saturation ratios, S. The *heavy lines* are for growth around a singly-charged ion, the *lighter lines* for uncharged (homogeneous) nucleation. There is an obvious, rapid decrease in barrier height when the saturation ratio, S, increases. The scale at the bottom gives the number of water molecules in the crystal at the indicated sizes

The existence of a large energy barrier is seen in Fig. 9.3; as the partial pressure of water vapour is increased, leading to a corresponding increase in the saturation, S, the energy barrier decreases. Note that the presence of a positive charge on the embryo not only decreases the height of the energy barrier slightly but confers stability on an embryo containing several water molecules which has a free energy somewhat larger than that of an uncharged (unstable) embryo of similar size. The sensitivity of the energy barrier to S, and thus indirectly to the temperature T, means that the statistical probability of an embryo having sufficient energy to pass the barrier (and reach the size at which steady growth follows) is extremely sensitive to temperature.

In any particular set of temperatures at the top of the mesosphere it follows that nucleation will occur effectively only at the height which shows the lowest temperature. But nucleation at a good rate does not necessarily mean that a visible cloud will form. Having passed the energy barrier and starting to grow, the infant cloud particle must survive long enough to grow to a size where it can scatter light effectively. This last condition means that there has to be a sufficiently large water vapour pressure for the particle to become large before it falls out of the saturated region. Therefore, to some extent the conditions for, first, nucleation and then growth are not related. One needs a low temperature to give high S for there to be nucleation; then the atmospheric layer immediately below the nucleating layer needs to be wet enough to allow the particles to grow. In summary, the nucleation occurs, if at all, at the mesopause. If the mesopause is broad, i.e. it consists of a layer of constant temperature rather than showing a distinct minimum in the temperature-height curve, nucleation will occur at the bottom of that layer, where S will be largest. The variation of S with height must always show a sharp maximum unless there is a pronounced increase in the water vapour mixing ratio with increase in height. This is unlikely to be the case.

A comprehensive review of nucleation of ice at low pressure and low temperature has been published recently by Roddy (1986). He places particular emphasis on the role of meteoric dust. An important consideration is that the nucleus should have a crystal type and lattice parameters similar to those of ice. The amount of misfit governs whether nucleation will be coherent (close match) or incoherent.

If the misfit is relatively large, lattice dislocations must occur near the interface between the nucleus and the sheath of ice. The dislocations will cause an increase in the free energy barrier to aggregation. If the misfit is small, elastic strain rather than dislocation will occur in the two lattices (the nucleus and the ice) and the increase, ΔF, in the free energy barrier is calculable. Clearly, the smaller the misfit, the smaller is ΔF; Roddy points out that meteoric material (such as MgO, $MgO\text{-}FeO$, $FeSi$) has misfit parameters less than 1%. For comparison, the popular cloud-seeding material, AgI, has a misfit parameter of 2.1%.

9.9 Settling of Particles

Once nucleated and over the free energy barrier, the infant particle grows and begins to fall at ever increasing speed as it grows. It falls, in most cases, through a region where the temperature is increasing in the downwards direction. The saturation, S, is therefore decreasing downwards and at some height S becomes equal to unity. Thereafter, the particle falls through increasingly unsaturated air and sublimation takes place quickly. The fall speed becomes smaller as the particle size becomes smaller; the shrinking particle lingers in the region just below the frost point level and the base of a noctilucent cloud must be very close to the level at which S is unity. If there were to be a precise measurement of temperature at the base of a noctilucent cloud, it would be possible to deduce with some reliability the water vapour mixing ratio at that time and place.

The cloud particles are not necessarily spherical; Reid (1975) has calculated the fall speeds of submicron particles, for both spheres and cylinders. He assumed specular reflection of atmospheric molecules from the particle surface. A better assumption is to allow for diffuse reflection with accomodation (Epstein 1924). Consequently, Reid's estimates of the fall speed should be decreased by about 40%. For the non-spherical particles, one should regard them as tumbling rapidly through the operation of Brownian movement (Einstein 1906).

Rosenfeld (1984, 1986) has recently raised an interesting point about the movement of cloud particles experiencing the passage of an internal gravity wave. With such a wave propagating energy from below, and moving obliquely through a noctilucent cloud layer, the individual cloud particles will experience a net *upward* drift. Rosenfeld calculated that this drift velocity can offset the downward (settling) speed of a cloud particle with a radius up to 0.2 to 0.3 μm.

If this occurred over a wide region, the result would be a stable cloud layer of monodisperse particles. There would be a single particle size at which the vertical component of the upward drift would exactly balance the settling speed. Smaller particles would be driven upward and into a rising temperature where they could no longer grow. Larger particles would simply fall out of the cloud and into a regime of rising temperature where they would evaporate. It seems likely that while this mechanism could indeed be effective on occasion, it is unlikely to have a major influence on the formation and existence of all noctilucent clouds. Only a fraction of the number of noctilucent clouds show wave motion to be present and those that do are not particularly noticeable and so presumably do not possess especially favourable conditions for particle growth.

A high-powered radar, operating at VHF, has been used for observation of the upper atmosphere over Poker Flat, Alaska ($\phi = 65°N$). The peak transmitted power exceeds 2 MW. An aerial array, 200×200 m in size, allows the study of incoherent scattering from the atmosphere up to altitudes in excess of 100 km. Balsley et al. (1983) reported the occurrence of a layer showing intense scattering that is present only during the summer months of May to August.

The layer is narrow (± 2 km half-power width). The curve of the signal:noise ratio, averaged over the period 18 June to 21 August 1980 shows a peak at 86.3 km, with the bottom of the scattering being at about 78 km and the top off the graph, above 99 km. The majority of individual measurements of the layer lie between 85 and 90 km. Balsley and colleagues (1983) presented persuasive arguments that this summertime scattering layer is produced by the shear instability of "tidal" winds in the upper atmosphere. Because of the deep stratification in temperature around the mesopause, the vertical wave number of low frequency periodicities in the wind field will be at a maximum just above the mesopause. There will be a rapid increase in wave number in just the last kilometre or two as the mesopause is reached from below.

The radar observations show a lower scattering layer (around 70-km height) to be present during the winter and *this* one, they feel, marks the level at which internal gravity waves break, causing turbulence. The lower layer is a wintertime phenomenon only, possibly because in summer upwardly propagated internal gravity waves are greatly attenuated in the stratosphere and mesosphere.

It is unavoidable to associate in one's thinking the presence of the summertime radio-scattering layer with the existence and occurrence of processes going on in the atmosphere that have a profound influence on the nucleation and growth of noctilucent clouds.

To turn to a completely separate phenomenon, noctilucent clouds are seen by scattering of sunlight. The cloud particles are therefore in intense, directed radiation and may show radiometer effects, or photophoresis, as discussed by Orr and Keng (1964) for stratospheric particles.

Sitarski and Kerker (1984) have made Monte Carlo calculations of photophoresis for a variety of particles at several gas pressures. In particular, they have included the case of an extremely low pressure, corresponding to the atmosphere at 92 km. For particles of 0.022 and 0.16 μm radii, the photophoretic forces are much smaller than the weights of the particles. The largest forces at this pressure are for particles of carbon or soot for which the photophoretic forces are 0.54 and 0.32 of the respective weights. For a dust particle, with a smaller refractive index and absorption coefficient, the photophoretic force is only 0.11 of the weight. The presumption is that for ice, and even for "dirty" ice, the photophoretic forces will be quite negligible compared with the particles' weights.

9.10 Modelling Noctilucent Clouds by Numerical Simulation

In the upper atmosphere, it is clear that many distinct effects occur at any one time, with all causing a noctilucent cloud to grow or to disappear.

McKay (1985) has directed attention to modelling the atmosphere for heights between 50 and 120 km. His investigation was specifically directed towards the occurrence of low temperatures during the summer. The model includes perforce chemical reactions, diffusion and eddy transport, and a detailed thermal balancing involving radiation, chemical potential energy, conduction, airglow emission and radiation in the infrared from carbon dioxide.

The temperature of the mesopause and the water vapour mixing ratio there (and just below) are probably the critical parameters in studying the formation and development of noctilucent clouds. The number density of nuclei, of both ions and meteor smoke, is important too, of course; Kashtanov and Novikov (1981) proposed that the chemical composition of the region is important, quite separately from the reactions that determine the proportion of hydrogenic material present as water molecules. Clearly, computer simulation, taking many processes into account, can lead to predictions showing much detail in both space and time. Gibbins et al. (1982) have drawn attention to the variability of water vapour mixing ratios (over the range 2–12 ppm) in the mesosphere.

In fact, what is observed is represented by a very limited set of data. Unless rocket sounding is used, the particle size distribution is inferred from measurements of radiance and polarization made for the whole cloud layer, from where nucleation is going on in the topmost parts down to the cloud base. The height of the cloud base may be known from two-station measurements but just precisely where this is on the temperature-height curve for the mesosphere and mesopause at that particular time is not usually known. In comparing predictions which are made from well-controlled theoretical analysis with actual observations it has to be remembered, therefore, that the model needs to be summed over all heights.

Some calculations based on the ideas set out in preceding sections and applied in a simple way are given by Gadsden (1981). In these calculations, estimates of particle size distributions are made which may prove helpful in assessing observational data. The characteristic of the size distributions is that they are broad, and extend from the smallest sizes up to a cut-off size (see Fig. 9.4). In the size range considered to be likely, the light scattering efficiency of a particle grows quickly with increasing particle radius. The light scattered from distributions such as those shown in Fig. 9.4 would be very much characteristic of a monodisperse distribution. The characteristic radius would be such as that indicated by the vertical lines above the upper end of each distribution. It seems likely that water vapour mixing ratios as high as 10^{-4} would be needed to provide cloud particles as large as 0.6 or 0.7 μm in radius.

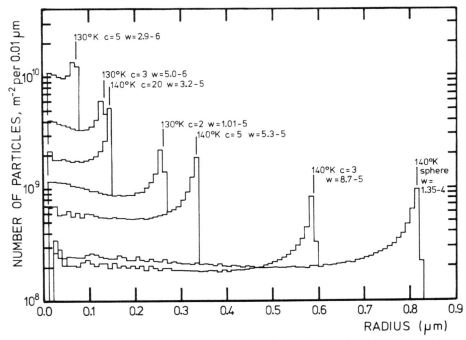

Fig. 9.4. Size distributions of typical noctilucent clouds for particular values of mesopause temperature and water vapour mixing ratio (w). The calculations have been made for ice cylinders with differing ratios, c, of length to radius. (The *lowest curve* is for a sphere to compare with the curve for 140 K, w = 8.7×10^{-5}, c = 3. The two curves have almost the same peak number of particles)

Shortly after these calculations were published, a major attempt to model noctilucent clouds was published by Turco and colleagues (Turco et al. 1982). A time-dependent, one-dimensional model has been programmed on a computer. It treats ice crytals, meteor dust, water vapour and ions as one interacting system. The calculations confirm that low temperatures ($<$ 140 K) at the mesopause is a sine qua non for noctilucent cloud nucleation. A water vapour mixing ratio of a few parts per million can lead to the formation of a visible noctilucent cloud. Clearly there may be, at smaller mixing ratios, a precursor of a noctilucent cloud that remains subvisual.

Turco et al. showed that growth onto *solid* nuclei (e.g. meteor smoke) may cause such rapid growth that nucleation onto ions is held back: in this case, the noctilucent cloud consists of a relatively small number of relatively large ice particles, with an optical thickness for the whole noctilucent cloud of approximately 10^{-4}. Without solid nuclei, or at very low temperatures ($<$ 130 K) growth on hydrated ions tends to lead to a greater number of smaller particles and an optical thickness of the order of 10^{-5}.

10 The Nature of Noctilucent Clouds

10.1 Introduction

In the century that has elapsed since systematic observations of noctilucent clouds began, many observational data have been recorded and discussed at length. From the very start of the period, hypotheses have been formed and, to the extent practical, tested. Quite naturally, because noctilucent clouds came to notice with the increased awareness of twilight sky phenomena following the Krakatoa eruption, the earliest suggestions were that noctilucent clouds were simple layers of volcanic dust appearing (after a delay of a year or two) in the uppermost parts of the atmosphere. These hypotheses later gave place to more complex explanations, to account for the clouds appearing at all in a part of the atmosphere that is very dry and at a very low pressure (cf. Chvostikov 1966b; Hesstvedt 1961, 1962, 1964, 1969b; Iwaskawa 1982; Turco et al. 1982; Vestine and Deirmendijan 1961; Webb 1965; Witt 1969).

10.2 Formation of Noctilucent Clouds

A theory of noctilucent clouds must first of all explain the sporadic appearance of noctilucent clouds, their absence during the months of September to March (in the northern hemisphere), their nearly constant height of about 82 km and their physical appearance on any particular occasion (cf. Charlson 1965; Soberman 1968).

After the eruption of Krakatoa in the Sunda Straits in 1883, notable twilight phenomena were seen all over the world (cf. Förster 1906; Kießling 1885; Pernter 1889 and the report by the Krakatoa committee of the Royal Society edited by Symmons 1888). It was soon conjectured that noctilucent clouds might be caused by volcanic eruptions because they appeared with increased frequency after other volcanic eruptions (see Table 10.1).

The conjecture that noctilucent clouds were caused by volcanic eruptions was rejected as early as 1926 by Jardetzky (1926) who suggested that noctilucent clouds were to be seen irrespective of earlier volcanic activity. Vestine (1934) suggested that there was an increase in the influx of extraterrestrial particles during the years 1880–1887 which could explain the high number of observations of noctilucent clouds compared with the number during the years 1910–1930. The increase in flux was supposed to come from the higher

Table 10.1. Period of enhanced NLCs and volcanic eruptions

Period of NLC	Volcanic eruption
1885–1889	Krakatoa, 1883
1908	Shtyubelya, Sapka, 1907
1911	Taal, 1911
1925–1929	Visokoi Island, 1922, 1929;
	Gunung Batur, 1925, 1926
1932–1938	Quizapa, 1932; Tzerrimai, 1937, 1938;
	Raluan, 1937; Ghai, 1937
1952–1960	Larrington, 1951;
	Bezymjannayam, 1955; Manam, 1956
1963–1968	Agung, 1963; Taal, 1965

frequency of great meteor showers and of bright comets in 1880–1910 than in 1910–1930. On the other hand, there were six "volcanic explosions" during 1880 to 1910 with more (nine) in the later period, 1910–1930.

However, it must be pointed out that since 1885, the number of noctilucent cloud displays reported per year may well fluctuate through changes in the amount of attention given to them by observers. It may be noted that e.g. in central Europe, opportunities for systematic noctilucent cloud observations were poor in the miserable aftermath of the destructive First World War (1914–1918).

Fogle and Haurwitz (1973) have discussed the long-term variations in the appearance of noctilucent clouds. They suggested that these variations might partly be caused by penetration of large amounts of water vapour into the stratosphere either from explosive volcanic events, or from such events as nuclear explosions. The two periods of increased noctilucent cloud activity (1885–1890 and 1963–1968) followed the two great volcanic eruptions of Krakatoa in 1883 and Agung in 1963. These eruptions were of the so-called Plinian type (cf. Ertel 1967) the events are violently explosive because plugged vents cause the buildup of large pressure with the conversion of magmatic water to steam. The efflux from Plinian-type eruptions is rich in water vapour, other gases and pumice. On the other hand, it is pertinent to point out here that the eruption of Katmai (58°N, 155°W) in 1912 did not produce increased noctilucent cloud activity.

With regard to their analysis Fogle and Haurwitz (1973) concluded:

We are well aware that our suggestion concerning the role of volcanic eruption of the Plinian type in the enhancement of noctilucent cloud activity is based at present on weak and partly even conflicting evidence and, as far as the atmospheric circulation is concerned, largely on surmise.

Thomas (unpublished work) from Boulder is the most recent adherent to the proposition that noctilucent clouds have essentially an origin in alterations of the atmosphere through volcanic injections of water, dust, etc. He points to the recent increase in numbers of volcanic eruptions in equatorial regions, the present secular increase in frequency of noctilucent clouds (Gadsden 1982) and an increase in hydroxyl airglow intensities which possibly indicates an increase in water vapour content of the upper atmosphere. He suggests that there exists an equatorial upwelling of the atmosphere which can provide a path from the lower stratosphere over the equator to the upper mesosphere in polar regions (cf. also Thomas et al. 1989).

10.3 Growth of Noctilucent Cloud Particles

The microphysics of growth of ice on a nucleus, based either on an ion or on a submicroscopic solid particle, has been discussed earlier (Sect. 9.7). Whether the nucleus is an ion or a solid particle seems to be of little importance in discussing the existence of a noctilucent cloud: there are more than enough nuclei present ab initio. The important questions are whether the cloud particles can grow in sufficient numbers to form a detectable, or visible, cloud; and to what size the particles are able to grow, given the atmospheric conditions expected in the neighbourhood of the mesopause in summer.

The computer model described by Turco et al. (1982; see Sect. 9.10) is being used by Jensen to model polar mesospheric clouds and to compare, with much success, the computed characteristics with the SME data. In as yet unpublished work, Jensen finds that spherical polar mesospheric cloud particles will have radii no great than 0.07 μm if the initial water vapour mixing ratio is 2 ppm by volume. This result is very similar to the 0.08 μm found by Gadsden (1981) in an earlier, and relatively simple, calculation using similar parameters. The distribution of sizes computed by Jensen is close to rectangular, that is to say with equal numbers in equal intervals of radius, up to a cutoff at some critical radius. Beyond this critical radius, the numbers decrease very quickly with increasing radius; at 0.1 μm, the computed particle numbers are down to 1 or 2% of those at radii less than 0.07 μm.

None of the models (Gadsden 1981b; Turco et al. 1982; Jensen et al. 1988) contain the effect of gravity waves passing through the level of a polar mesospheric cloud or a noctilucent cloud to provide an upward component of acceleration on the cloud particles. Rosenfeld, in a series of papers (Rosenfeld 1984), has proposed that a gravity wave with representative amplitude and frequency can support solid particles (with an assumed bulk density equal to 3000 kg m^{-3}) of radius 0.2 μm. If the particle density be taken instead as 930 kg m^{-3} (solid ice at 130 K), then particles of radius up to 0.65 μm will be supported. If the cloud particles were to have a loose structure or to be non-spherical, with either or both characteristic resulting perhaps from aggregating water molecules at low pressure and exceedingly low temperatures, then the critical radius, that is, the radius at the cut off in numbers, will be of the order of 1 μm.

10.4 Evaporation of Noctilucent Cloud Particles

Irrespective of the details of the processes by which cloud particles grow to their maximum size, eventually they will fall into the unsaturated region of the atmosphere lying below the cloud layer. The atmosphere below a polar mesospheric cloud or a noctilucent cloud is very likely to exhibit a temperature monotonically falling over a considerable range of increasing heights. The falling particle, therefore, experiences surroundings which are rising in temperature (cf. Thomas and McKay 1985, 1986).

At the very low temperatures experienced by a cloud particle, it is not the relative humidity or atmospheric saturation ratio that is important for evaporation, but the absolute temperature. In the low pressure conditions around the particle, the net outward flux from the particle surface is controlled by the surface temperature of the particle. As the particle falls into surroundings with ever-increasing temperature, its surface must warm up until, at 150 K or higher, the rate of evaporation (a rapidly increasing function of temperature) becomes large enough that even a 1-μm particle evaporates in a few seconds. Thus, the lower border of a polar mesospheric cloud or a noctilucent cloud marks a particular isothermal surface (approximately 150 K) in the atmosphere rather than the 100% humidity level. Certainly all models agree in putting the cloud base well below the mesopause level: it is incorrect, therefore, to speak of noctilucent clouds being "at the mesopause".

Alternatively, the base of a noctilucent cloud may mark the region in which upwardly propagating gravity waves break. There will be, in this region, much turbulence which will put a step into the profile of atmospheric temperature. A cloud particle, as it falls into the top of the turbulent region, will experience a sudden jump in temperature. Naturally, in this case (since the gravity waves are breaking at or below the cloud base), gravity wave support during the growth of a cloud particle is not available to allow enhancement of particle sizes over what would be expected in an undisturbed atmosphere.

10.5 The Relationship Between Polar Mesospheric Clouds and Noctilucent Clouds

At a recent workshop, Gadsden presented an analysis which brought together SME satellite data provided by Jensen and Clancy in Boulder and the NW European observations collected by Gavine in Edinburgh. The analysis shows that the southern borders of noctilucent clouds lie typically some 10° of latitude to the south of the southern borders of polar mesospheric clouds in the same longitude sector. Furthermore, although there is a loose positive association between the occurrence of the brighter polar mesospheric clouds being followed at night by brighter or more extensive noctilucent clouds, the association is very loose and contrary examples of both kinds exist, and are not rare (cf. Schröder 1987, 1988b).

Nothing in the comparison between the occurrence of polar mesospheric clouds well to the north of the European land mass and the appearance of noctilucent clouds over NW Europe contradicts the idea that polar mesospheric clouds are the nursery of noctilucent clouds. Just as children do not necessarily grow up to be all that their parents wish, so polar mesospheric clouds may not develop within a few hours or a day or two into noctilucent clouds. Given the general equatorward flow of air at these altitudes and latitudes during the summer, seeing noctilucent clouds at lower latitudes than the polar mesospheric clouds leads naturally to the working hypothesis that a noctilucent cloud represents the ultimate in growth of a small proportion of polar mesospheric cloud particles. Perhaps polar mesospheric clouds should be regarded as a sperm bank: some small proportion, through chance, develop into large bodies (cf. Gadsden and Schröder 1989; Schröder 1987, 1988b).

If this is indeed the morphological picture, then polar mesospheric clouds result from the chilling of upwelling stratospheric air over polar regions. For the reasons examined closely in the models of Turco et al. (1982) and Jensen et al., the particle sizes in a dense polar mesospheric cloud cannot exceed 0.07 μm and the polar mesospheric clouds will be relatively unobtrusive in the twilight sky, as seen by an observer on the ground. Indeed, the entire life of a polar mesospheric cloud may be confined to the mesosphere lying above, or just to the north of, sites where the twilight sky does not become dark enough to allow the polar mesospheric clouds to be seen by an observer on the ground.

Suppose that at times conditions occur when there are gravity waves or when rather larger amounts of water vapour occur in the mesosphere. Some (perhaps one in a hundred) of the polar mesospheric cloud particles will grow to submicron size, taking several hours or a day or two to achieve this size. During the period of growth, they will be swept equatorwards by the prevailing flow and become large enough to form a visible noctilucent cloud. This noctilucent cloud may well show structure as gravity waves pass through the layer, or be in a region of much wind shear where locally excited waves may be present. In either case, the presence of the structure brings the noctilucent cloud to the attention of a visual observer looking up into the twilight sky which, because of his lower latitude, is dark enough to make the noctilucent cloud visible.

The equatorward transport of cloud particles made up of major assemblies of water molecules, with the evaporation at a well-defined cloud boundary and at a well-defined height must have a detectable effect on the chemistry of the region just in the vicinity of the cloud base. Perhaps this accounts for Shefov's observations of the hydroxyl nightglow, discussed above in Section 8.2.

10.6 Summary

Recently, interest in noctilucent clouds has grown markedly: international workshops are occurring at an accelerating pace and this final chapter has been written after the workshop held in Boulder (USA) in March 1988, and before the workshop to be held in Tallinn (ESSR) in July 1988. (The Proceedings of the Boulder workshop are to be published in the Journal of Geophysical Research — Oceans and Atmospheres — while those of the Tallinn workshop will probably appear as a separate publication of the Academy of Sciences of the Estonian SSR.)

It is clear that current thinking about noctilucent clouds accepts that the cloud particles are at the large-radius end of a distribution of sizes of clustered water molecules. The interesting points in noctilucent cloud theory seem to be:

1. How large can the particles grow;
2. What is the morphology of the relation between polar mesospheric clouds and noctilucent clouds;
3. Modification of the (gravitational) settling rate of cloud particles in polar mesospheric clouds and/or noctilucent clouds through the breaking of gravity waves and whether this is necessary to the growth of particles;
4. What chemical and ionic effects result from the aggregation of extremely large numbers of water molecules in clusters with the consequent release in a limited volume of significant numbers of water molecules;
5. Possible electrodynamic effects associated with polar mesospheric clouds and noctilucent clouds; and, lastly,
6. The link between occurrence of polar mesospheric clouds and noctilucent clouds, in the upper atmosphere, with the meteorology of the troposphere and with changes in the Earth's climate.

Certainly, the mesopause region of the atmosphere is but poorly understood. There are major discrepancies between numerical models and the observed data. Noctilucent clouds offer a naturally occurring tracer for use in studies of waves, winds and settling processes, for use in estimating the transport of water vapour and dust in the atmosphere, regardless of whether the dust is terrestrial or extraterrestrial in origin. It may prove to be that noctilucent clouds are also a sensitive indicator of approaching major alterations in the global climate. Only time, and further research, will tell (cf. Schröder 1968a,b,d, 1971; Thomas et al. 1989).

11 Bibliography

A) Before 1900

Arago F (1838) Ann Bur des Longitudes – Sci Notes sur le Tonnere, 279–283, Paris

Arago F (1855) Storm clouds ever continuously luminous. Meteorol Essays, Longman, London 48–52

Archenhold FS (1894) Die Verwendung der Photographie bei der Erforschung der leuchtenden Nachtwolken, Sternschnuppen und Nebelflecken. Photogr Mitt 31:3–6

Backhouse TW (1885a) The luminous cirrus cloud of June and July. Meteorol Mag 20:133

Backhouse TW (1886a) Luminous clouds. Nature (London) 42:246

Backhouse TW (1886b) The bright clouds and the aurora. Nature (London) 34:386–387

Backhouse TW (1886c) Lett R Soc (Arch)

Backhouse TW (1887a) The sky-coloured clouds. Nature (London) 36:269

Backhouse TW (1887b) The sky-coloured clouds. Nature (London) 42:246

Battermann H (1886) Bemerkungen zu dem Aufsatz des Herrn O Jesse 'Die auffallenden Abenderscheinungen am Himmel im Juni und Juli 1885'. Meteorol Z 3:179–180

Deluce F (1787) Idee sur la météorologie, tome 2. Veuve Duchesne, Paris

Förster W (1892) Beobachtungen der leuchtenden Nachtwolken. Mitt Ver Freunde Astron Kosm Phys 2:80

Hartwig E (1893) Beobachtung der leuchtenden (silbernen) Nachtwolken zu Dorpat und Bamberg. Ber Naturforsch Ges Bamberg 16:1–4

Helmholtz Hv (1888) Über atmosphärische Bewegungen. Sitz Ber Akad Wiss Berlin, S. 647–663

Helmholtz Hv (1888b) Über atmosphärische Bewegungen (Zweite Mittheilung) Sitz Ber Akad Wiss Berlin, S. 761–780

Helmholtz Rv (1887) Silberne Wolken. Meteorol Z 4:335

Jesse O (1885) Auffallende Abenderscheinungen am Himmel. Meteorol Z 2:311–312

Jesse O (1886a) Die auffallenden Abenderscheinungen am Himmel im Juni und Juli 1885. Meteorol Z 2:6

Jesse O (1886b) Silberne Wolken am nächtlichen Himmel. Meteorol Z 2:356

Jesse O (1887a) Die Beobachtungen der leuchtenden Nachtwolken. Meteorol Z 4:179–181

Jesse O (1887b) Die Höhe der leuchtenden (silbernen) Wolken. Meteorol Z 4:424

Jesse O (1888) Über die leuchtenden (silbernen) Wolken. Meteorol Z 5:90–94

Jesse O (1889) Von den leuchtenden Nachtwolken. Himmel u Erde 1:428–429

Jesse O (1890a) Anweisungen für die photographischen Aufnahmen der leuchtenden Nachtwolken. Sonderdr Verein Freunde Astron Kosm Physik. Berlin (o.V.)

Jesse O (1890b) Untersuchungen über die sogenannten leuchtenden Wolken. Sitzungsber Preuss Akad Wiss Math-Physik. 1890:1033–1035

Jesse O (1891) Über die leuchtenden Nachtwolken. Meteorol Z 8:306–308

Jesse O (1896) Die Höhe der leuchtenden Nachtwolken. Astron Nachr 140:161–168

Kepler J (1604) Ad Vitellionem paralipomena, quibus astronomiae pars optica traditur, Marnium and Aubri, Frankfurt/M

Kießling J (1885) Die Dämmerungserscheinungen im Jahre 1883 und ihre physikalische Erklärung. Voss, Hamburg/Leipzig

Kießling J (1888) Untersuchungen über Dämmerungserscheinungen zur Erklärung der nach dem Krakatau-Ausbruch beobachteten atmosphärisch-optischen Störung. Voss, Hamburg/Leipzig
Le Conte J (1889) Noctilucent clouds. Nature (London) 32:245
Leslie R (1885) Sky glows. Nature (London) 32:245
Leslie RC (1886) Luminous clouds. Nature (London) 34:264
Leslie RC (1892) Luminous clouds. Nature (London) 46:296
Mairan JJ (1733) Traité physique et historique de l'aurore boréale. Acad Royal, Paris
Maignon E (1648) Perspectiva horaria sive horographia gnomonica tum theoretica, tum practica libri quator. Robeus, Roma
Marron y Miranda, Manuel M (1899) El catorce de Noviembro. Mexico: 30
Omond RT (1888) 'Sky coloured' clouds at night. Nature (London) 38:220
Penning E (1888) Die leuchtenden Wolken. Meteorol Z 5:368-369
Pernter JM (1889) Der Krakatau-Ausbruch und seine Folge-Erscheinungen (4: Die optischen Erscheinungen). Meteorol Z: 6:447-466
Peters CFW (1894) Lehrbuch der kosmischen Physik. Vieweg, Braunschweig
Pokrovskii KD (1897) Luminous clouds. Izv Russ Astron Obs 6:273-289 (in Russian)
Pröll G (1889) Leuchtende Nachtwolken. Meteorol Z 6:478
Rayleigh Lord (1881) On the electromagnetic theory of light. Philos Mag (5th Ser) 12:81-101
Rowan DJ (1886a) Luminous boreal clouds. Nature (London) 34:192
Rowan DJ (1886b) Luminous clouds. Nature (London) 34:264
Rowan DJ (1887) Luminous boreal cloudlets. Nature (London) 36:245
Rowan DJ (1889) Luminous night clouds. Nature (London) 40:151
Rowan DJ (1890a) A remarkable appearance in the sky. Nature (London) 42:222
Rowan DJ (1890b) The night-shining clouds. Nature (London) 42:246
Schmidt J (1869) Publ Sternwarte Athen I Ser II:1-40
Shaw C (1890) The night-shining clouds. Nature (London) 42:246
Smyth CP (1883) Bright clouds on a dark night sky. Trans R Soc Edinburgh 32:11
Smyth CP (1886) The silver-blue cloudlets again. Nature (London) 34:311-312
Sreznevskii B (1897) The height of noctilucent clouds. Meteorol Vestn 7:130 (in Russian)
Symmons GJ (1888) The eruption of Krakatoa and subsequent phenomena. Rep Krakatoa Comm R Soc London
Tseraskii VK (1890) Luminous clouds. Ann Observ Moscow 2:177 (in Russian)
Verbeek RDM (1885) Krakatau. Batavia
Verdeil F (1783) Mem Soc Sci Phys Lausanne 1:110

B) 1900-1950

Archenhold FS (1928) Die leuchtenden Nachtwolken und bisher unveröffentlichte Messungen ihrer Geschwindigkeit. Weltall 27:137-144
Astapovitsch IS (1939) Noctilucent clouds. Izv Akad Nauk SSSR Ser Geogr Geofiz 2:183-204 (in Russian)
Barkow E (1916) Über eine Beobachtung von selbstleuchtenden Wolken. Meteorol Z 33:469
Battermann H (1911) Leuchtende Nachtwolken. Meteorol Z 28:416
Bauernberger H (1900) Leuchtende Nachtwolken zu Linz, 28. Juni 1900. Meteorol Z 17:419
Bronshten VA and Zatejssikov (1938) Noctilucent clouds in 1936. Meteorol Gidrol 4:3-24 (in Russian)
Busch F (1908) Leuchtende Nachtwolken am Nordhorizont. Meteorol Z 25:314
Chapman S (1931) Some phenomena of the upper atmosphere (Bakerian Lecture) Proc R Soc London Ser A 132:353-374
Einstein A (1906) Zur Theorie der Brownschen Bewegung. Ann Phys Ser 4 19:371-381
Epstein PS (1924) On the resistance experienced by spheres in their motion through gases. Phys Rev 23:710-733
Ertel H (1938) Methoden und Probleme der dynamischen Meteorologie. Springer, Berlin

Ertel H (1939) Die theoretischen Grundlagen der dynamischen Meteorologie. Meteorol Taschenb,
    Fünfte Ausg, hrsg. v. F. Linke, Akademische Verlagsges, Leipzig: pp 1–33
Fedynsky VV (1939) The single-station method for processing noctilucent clouds. Astron Zh
    16:42–46 (in Russian)
Fessenkov VG (1949) The mass of the atmospheric residue of the Sikhote-Alin meteorite. Dok Akad
    Nauk SSSR 66 (in Russian)
Foerster W (1900) Die leuchtenden Nachtwolken. Himmelswelt 10:66
Foerster W (1906) Von der Erdatmosphäre zum Himmelsraum. Hillger, Berlin/Leipzig
Foerster W (1908) Mitteilung, betreffend des Erscheinens der sogenannten leuchtenden Wolken.
    Himmelswelt 18:62–70
Foerster W (1911a) Lebenserinnerungen und Lebenshoffnungen. G Reimer, Berlin
Foerster W (1911b) Zur Frage des widerstehenden Mittels. Meteorol Z 28:332–333
Garfinkel B (1944) Astron J 50:169–179
Gruner P, Kleinert H (1927) Die Dämmerungserscheinungen. Grand, Hamburg
Hartmann W (1933) Über leuchtende Nachtwolken in Deutschland im Juli 1932. Meteorol Z
    50:272–273
Haurwitz B (1931) Zur Theorie der Wellenbewegungen in Luft und Wasser. Habil.-Schrift
    Universität Leipzig, 106 p
Hoffmeister C (1946) Die Strömungen der Atmosphäre in 120 km Höhe. Z Meteorol 2:33–41
Humphreys W J (1933) Nacreous and noctilucent clouds. Mon Weather Rev 61:228–229
Jardetzky W (1926) Über die leuchtenden Nachtwolken. Meteorol Z 43:310–312
Malzev V (1926) Noctilucent clouds on the night of 8/9 August 1925. Mirov 15:153–172
Malzev V (1929) Luminous night clouds. Byull Geof Fenol Rolm No 43:56–57
Mie G (1908) Beiträge zur Optik trüber Medien, spezielle kolloidale Metallösungen. Ann Phys Ser
    4 25:377–345
Paton J (1949) Luminous night clouds. Meteorol Mag 78:354–357
Peppler W (1935) Über den Ursprung der leuchtenden Wolken. Z Angew Meteorol 52:334–337
Rudzki MP (1908) Nordschein am 30. Juni 1908 in Krakau. Meteorol Z 25:313–314
Schmidt A (1908) Über die Leuchterscheinung in der Nacht vom 30. Juni auf den 1. Juli 1908.
    Himmelswelt 18:66
Scultetus HR (1934) Beobachtungen leuchtender Nachtwolken im Sommer 1933. Z Angew Me-
    teorol 51:264–266
Shinjo S (1914) Meteoreinfälle als Ursache des vermuteten Zurückhaltens der obersten Atmos-
    phäre. Mem Coll Sci Eng 1914:1
Stenzel A (1909) Die Dämmerungsanomalien im Sommer 1908. Meteorol Z 26:437–446
Störmer C (1925) Aus den Tiefen des Weltraums bis ins Innere der Atome. Brockhaus, Leipzig
Störmer C (1931) Merkwürdige Wolken im Höhenintervall 23–26 Kilometer über der Erde.
    Gerlands Beitr Geophys 32:63–68
Störmer C (1933) Height and velocity of luminous night clouds observed in Norway, 1932. Oslo Univ
    Publ 6:1–45
Störmer C (1935) Measurements of luminous night clouds in Norway 1933 and 1934. Astrophys
    Norv 1:87–114
Süring R (1943) Die Wolken. Akadem Verlagsges, Leipzig
Vassy A (1941) Ann Phys (11e Ser) 16:145–203
Vegard L (1933) Investigations of the auroral spectrum. Geofys Publ 10:53
Vegard L (1940) Continued investigations on the auroral luminescence and the upper atmosphere.
    Geofys Publ 12:14
Vestine EH (1934) Noctilucent clouds. JR Astron Soc 28:249–272 and 303–317
Wegener A (1912) Die Erforschung der obersten Atmosphärenschichten. Gerlands Beitr Geophys
    11:102
Wegener A (1925) Die Temperatur der obersten Atmosphärenschichten Meteorol Z 42:402–405
Wegener A, Wegener K (1935) Vorlesungen über die Physik der Atmosphäre. Akadem Verlagsges,
    Leipzig
Wörner H (1935) Beobachtung leuchtender Nachtwolken. Meteorol Z 52:379
Wolf M (1908) Über die Leuchterscheinung in der Nacht vom 30. Juni auf den 1. Juli 1908.
    Himmelswelt 18:68–70

Since 1950

Allen CW (1963) Astrophysical quantities, 2nd ed. Athlone, London

Anderson JG, Donahue TM (1975) The neutral composition of the stratosphere and mesosphere. J atmos terr phys 37:865–884

Andrews G, Holton JR, Leovy CB (1987) Middle atmosphere dynamics. Academic Press, Orlando

Apruzese JP, Schoeberl MR, Strobel DF (1982) Parametrization of IR cooling in a middle atmosphere dynamic model. J Geophys Res 87:8951–8966

Arakawa H (1958a) Luminous clouds. Temmon to kisjo 24:6 (in Japanese)

Arakawa H (1958b) On the influence of luminous clouds upon weather in the summer of Japan. Tenki 5:179 (in Japanese)

Arnold F (1980) Ion-induced nucleation of atmospheric water vapor at the mesosphere. Planet Space Sci Rev 28:1003–1009

Arnold F (1981) Solvated electrons in the upper atmosphere. Nature (London) 294:732–733

Arnold F, Joos W (1979) Rapid growth of atmospheric cluster ions at the cold mesopause. Geophys Res Lett 6:763–766

Arnold F, Krankowsky D (1971) Negative ions in the lower ionosphere: a comparison of a model computation and mass spectrometric measurements. J Atmos Terr Phys 33:1693–1702

Arnold F, Krankowsky D (1977) Water concentrations at the mesopause. Nature (London) 268:218–219

Asano S, Yamamoto G (1975) Light scattering by a spheroidal particle. Appl Opt 14:29–49

Astapowitsch IS (1959) Meteoric phenomena in the earth's atmosphere. Akad Nauk, Moscow (in Russian)

Astapowitsch IS (1959) Svodka nabludii serebristych oblakov v Rossii i v SSSR s 1885 do 1944. Trudy VI soves po sereb oblakam, Riga 1959 pp 49–92 (in Russian)

Austin J (1983) Krakatoa sunsets. Weather 38:226–231

Avakyan S, Avaste O, Willmann Ch, Kovalyonok V, Lazarev A, Savinykh S (1986) Noctilucent cloud observations in the equatorial and low latitudes carried out by main team of the 5th expedition of the orbital station 'Salyut-6'. Collect Works Int Workshop Noctilucent Clouds, Tallinn 1986 pp 121–130

Avakyan EV, Avaste O, Willmann Ch, Germashevsky M, Ivanchenkov AS, Klimuk PI, Kovalyonok VV, Lazarev AI (1981) Atmos-opt yablen nabluch orbit nauch stanchii 'Salyut'. Tartu, pp 52–57

Avaste O, Fedynsky AV, Grechko GM, Sevastyanov VI, Willman Ch (1980) Advances in noctilucent cloud research in the space era. Pageoph 118:528–580

Avaste O, Alekseev AM, Veismann UK, Willman Ch, Klimuk PI, Koksharov II, Lazarev AI, Sevastyanov VI, Sergevich VN, Fedorova EO, Eerme KA (1977) Opt Issl Izpuch Atmos Pol Siyanii i Serebr Oblakov Borta Orbit Nauch Statch 'Salyut-4', Tartu, pp 67–78 (in Russian)

Avaste OA, Willmann Ch I, Grechko GM, Romanenko YuV (1981) Atmosf-Optich yablen nabluch orbit nauch sachnii 'Salyut'. Tartu, pp 139–146 (in Russian)

Avaste O, Gadsden M, Grechko GM (1988) The coloured edge of noctilucent clouds. J Atmos Terr Phys 50:591–599

Avaste O, Keevallik SH (1982) Remote monitoring of aerosols from space. Adv Space Res 2:87–93

Baggaley WJ (1977) The meteoric night-glow. Mon Not R Astron Soc 181:203–210

Baibulatov F Kh, Ivaniya SP, Pas'ko LN (1971) Calculations of light scattering on spherical particles. Izv SO AN SSSR Ser Tekh Nauk 3, No 1

Baibulatov F Kh, Ivaniya SP (1976) Numerical studies of aerosol particle temperature in the upper atmosphere. Izv Atmos Ocean Phy 12:523–530

Baibulatov F Kh (1977) Izv Atmos Okean Fiz 13:1212–1214 (in Russian)

Balsley BB, Ecklund WL, Fritt DC (1983) VHF echoes from the high-latitude mesosphere and lower thermosphere: Observations and interpretation. J Atmos Sci 40:2451–2466

Banderman LW, Kemp JC (1973) Circular polarization by single-scattering of unpolarized light from loss-less, nonspherical particles. Mon Not R Astron Soc 162:367–377

Banks PM (1977) Observations of joule and particle heating in the auroral zone. J Atmos Terr Phys 39:179–193

Banks PM, Kockarts G (1973) Aeronomy. Academic Press, New York, London

Barat J (1973) Les mesures de turbulence. La turbopause. Rech Spat 12:20

Barber P, Yeh C (1975) Scattering of electromagnetic waves by arbitrarily shaped dielectric bodies. Appl Opt 14:2864–2872

Bary de E, Rössler F (1974) Particle distribution in noctilucent clouds. Beitr Phys Atmos 47:261–268

Batten ES (1961) Wind systems in the mesosphere and lower ionosphere. J Meteorol 18:283–291

Bavilaqua RM, Olivero JJ, Schwartz RP, Gibbons CJ, Bologna JM, Thacker DJ (1983) An observational study of water vapor in the mid-latitude mesosphere using ground-based microwave techniques. J Geophy Res 88:8523–8534

Belton MJS (1966) Dynamics of interplanetary dust. Science 151:43

Belyaev BI, Grechko GM, Gubarev AA, Kisevsky LI, Klimuk PI, Koksharov II, Lovchikova LP, Plyuta VE, Sevastyanov VI, Sergeyevich NV, Yanovsky AF (1979a) Issl Atmos Optich Yavch Borta Orbit Nauch Stanch 'Salyut-4', Tartu, pp 167–182 (in Russian)

Belyaev BI, Kiselevsky LI, Klimuk PI, Lovchikova LP, Plyuta VI, Sevastyanov VI, Yanovsky AF (1981) Atmos Optich Yablen Nabluch Orbit Nauch Stanchii 'Salyut', Tartu, pp 58–69 (in Russian)

Belyaev BI, Kiselevsky LI, Klimuk PI, Koksharov II, Lovchikova LP, Plyuta VE, Sevastyanov VI, Sergeyevich VN (1979b) Spectroscopic investigation of silver clouds in the visible region of the spectrum from the orbital scientific station 'Salyut-4'. J Appl Spectrosc 30:471–474

Benech B, Dessens J (1974) Mid-latitude artificial noctilucent clouds initiated by high-altitude rockets. J Geoph Res 79:1299–1301

Bernhardt KH (1982) Leuchtende Nachtwolken Z Meteorol 33:252

Bessonova TD (1963) Apparent frequency of noctilucent clouds appearance according to observations carried out by the USSR Hydrometeorological service network of stations for 1957–1959. Issl Meteorol 6:62–63 (in Russian)

Bevilacque RM, Wilson WJ, Ricketts WB, Schwartz PR, Howard RJ (1985) Possible seasonal variability of mesospheric water vapor. Geophys R Lett 12:379–400

Bezrukova AY (1967) Some considerations concerning reaction between occurrence of noctilucent clouds with increase in zonality of atmospheric streams in the earth and the sunspot activity. Nablud serebrist oblakov, Nauka, Moscow, pp 79–83 (in Russian)

Biermann LW, Kemp JC (1973) Circular polarization by single scattering of unpolarized light from loss-less, non-spherical particles. Mon Not R Astron Soc 162:367–377

Björn LG (1984) The cold summer mesopause. Adv Space Res 4:145–151

Björn LG, Arnold F (1981) Mass spectrometric detection of precondensation nuclei at the arctic summer mesopause. Geophys Res Lett 8:1167–1170

Björn LG, Kopp E, Herrmann U, Eberhardt D, Dickinson PHG, Mackinnon DJ, Arnold F, Witt G, Lundin A, Jenkins DB Heavy ionospheric ions in the formation process of noctilucent clouds. J Geophys Res 90:7985–7998

Blamont JE, de Jager C (1961) Upper atmospheric turbulence near the 100 km level. Ann Geophys 17:134–144

Böhme W (1969) Über den zweijährigen Zyklus der allgemeinen Zirkulation und seine Ursachen. Geod Geophys Varöff NKGG Reihe II, No 9

Bohren CF (1983) On the size, shape and orientation of noctilucent particles. Tellus 35B: 65–72

Borbély E (1972) A meteorologiai elemek mezönh tavaszi és öszi atvelt ödasá ástartosférában. Idöjárás (Budapest) 76:351–363

Brigg EK (1953) The formation of atmospheric ice crystals by the freezing of douplets. QJR Meteorol Soc 79:510–519

Bronshten VA (1970) K istorii otkrytija i pervych issledovanija serebristach oblakov. Fiz mezsosf (serebr) obl. Riga pp 143–152 (in Russian)

Bronshten VA, Grishin NI (1970) Serebristye oblaka. Nauka, Moscow (in Russian)

Brown TJ (1973) The chemistry of metallic elements in the atmosphere and ionosphere. Chem Rev 73:645

Burov MI (1959) Methode photogrammetrique pour la determination de l'altitude des nuages argentes. Trudy sovesc po serebr oblakam, Tartu pp 92–111 (in Russian)

Burov MI (1966) Determination of parameters of noctilucent clouds in 1964. Meteorol Issled 12:33–46 (in Russian)

Burov MI (1967) Spatial and dynamics characteristics of noctilucent clouds. Int Symp, Tallinn, pp 200–207

Byrne FD (1964) Noctilucent clouds and aurora. Meteorol Mag 93:121

Byrne D (1964) Typescript (undated) found in papers belonging to the late James Paton

Chanin ML (1984) Review of lidor contribution to the description and understanding of the middle atmosphere. J Atmos Terr Phys 46:973–993

Chao J (1965) On the nature and shape of noctilucent clouds particles. M.S. Thesis University Alaska

Chapman S, Kendall PC (1965) Noctilucent clouds and thermospheric dust: their diffusion and height distribution. QJR Meteorol Soc 91:115–131

Chapman S, Kendall PC (1966) Noctilucent clouds and thermospheric dust: discussion. QJR Meteorol Soc 92:311–313

Charlson RJ (1965) Noctilucent clouds: a steady state model. QJR Meteorol Soc 91:517

Charney JG, Drazin PG (1961) Propagation of planetary scale distributions from the lower into upper atmosphere. J Geophys Res 66:83–109

Christie AD (1966) A note on recent noctilucent cloud observations over Canada. J Atmos Sci 23:446–449

Christie AD (1969a) The genesis and distribution of noctilucent clouds. J Atmos Sci 26:168–176

Christie AD (1969b) A condensation model of noctilucent cloud formation. Space Res IX:175–182

Chvostikov IA (1952) Serebristye oblaka. Priroda 41:49–59 (in Russian)

Chvostikov IA (1959) Temperature regime of the atmosphere in the region of noctilucent clouds. Trudy sovesc po serebrist oblakam Tartu pp 85–91 (in Russian)

Chvostikov IA (1961) Die Natur der leuchtenden Nachtwolken und die Temperatur der Atmosphäre in der Mesopause. Trudy VI sovesc po serebrist oblakam, Riga pp 7–12 (Russian)

Chvostikov IA (1963) Physics of the ozonsphere and of the atmosphere. Akad Nauk, Moscow (in Russian)

Chvostikov IA (1966a) Noctilucent clouds. Priroda 55:48–53 (in Russian)

Chvostikov IA (1966b) Nature of noctilucent clouds and some questions of photochemistry and dynamics of the earth's atmosphere. Meteorol Issl 12:57–65 (in Russian)

Chvostikov IA, Megrelishvilli G (1970) Atomic hydrogen, oxygen and molecules of H_2O in the upper layers of the atmosphere and some problems of geophysics and physics of extraterrestrial space. Fizika mezosf (serebrist) oblakov Riga 1970, pp 9–24 (in Russian)

Clairemidi J, Hersé M, Moreels G (1985) Bi-dimensional observation of waves near the mesopause at auroral latitudes. Planet Space Sci 33:1013–1022

Clarke D, Grainger JF (1971) Polarized light and optical measurement. Pergamon, Oxford

COSPAR (1965) CIRA 1965, COSPAR International Reference atmosphere (1965) North-Holland, Amsterdam

Currie BW (1963) Widespread occurrence of noctilucent clouds over Canada, June 26 to July 6. Can J Phys 41:1745–1746

Danilov AD (1984) Adv Space Res 4:67–71

Deguchi S, Muhleman DO (1982) Mesospheric water vapor. J Geophys Res 87:1343–1346

Deirmendijan D, Vestine EH (1959) Some remarks on the nature and origin of noctilucent cloud particles. Planet Space Sci 1:146–153

Dickinson RE (1975) Meteorology of the upper atmosphere. Rev Geophys Space Phys 13:771–790

Dieminger W (1973) Kopplungserscheinungen zwischen unterer und oberer Atmosphäre. Bonn Meteorol Abh 17:253

Dietze G (1973) Der Einfluß atmosphärischer und physiologischer Bedingungen auf die Homogenität von Beobachtungen leuchtender Nachtwolken. Gerlands Beitr Geophys 82:175–186

Dirikis MA (1959) Resultate der Beobachtungen der leuchtenden Nachtwolken in der Rigaer Abteilung der Allgemeinen astronomischen geodätischen Gesellschaft im Jahre 1958. Trudy sovesc po serebrist oblakam Tartu, pp 23–24 (in Russian)

Dirikis MA, Mookins EE (1966) Determination of the projection of noctilucent clouds on the earth's surface. Meteorol Issl 12:52–56 (in Russian)

Dirikis MA, Evdokimenko SV, Francmann JL (1966) Determination of space coordinates of noctilucent clouds. Meteorol Issl 12:47–51 (in Russian)

Donahue TM, Guenther B, Blamont JE (1972) Noctilucent clouds in daytime circumpolar particulate layers near the summer mesopause. J Atmos Sci 30:515–517

Donahue TM, Guenther B (1973) The altitude of the scattering layer near the mesopause over the summer pole. J Atmos Sci 30:515–517

Dubin M (1986) Nuctilucent clouds the problem of origin. EOS 67:319

Ebel A (1973) Jahreszeitliche Änderung der großräumigen Zirkulation im Bereich der Mesopause und unteren Thermosphäre. Ann Meteorol NF 6:229–231

Ebel A (1974a) Heat and momentum sources of mean circulation of an altitude of 70 to 100 km. Tellus 26:325–333

Ebel A (1974b) Dynamic processes as derived from the mass mean circulation in the upper mesosphere and lower thermosphere. Space Res XIV:195

Ebel A (1984) Zirkulation und Transport in der mittleren Atmosphäre, Mesosphäre und untere Thermosphäre. Mitt Univ Köln 39:92–98

Ebel A, Jakobs HJ, Speth P (1983) Turbulent heating and cooling of the mesopause and their parametrization. Ann Geophys 1:359–370

Eckart C (1960) Hydrodynamics of ocean and atmosphere. Pergamon, Oxford

Eisenberg D, Kauzmann H (1969) The structure and properties of water. Clarendon, Oxford

Ertel H (1953a) Der Symmetriecharakter des Turbulenz (Austausch) Tensors. Sitz Ber Dt Akad Wiss Berlin, Kl Math allg Naturwiss No 2:1–8

Ertel H (1953b) Entwicklungsphasen der Geophysik. Akademie-Verlag, Berlin

Ertel H (1967) Distribution of inhomogenous volcanic particles through turbulent diffusion. Gerlands Beitr Geophys 76:406–412

Farlow NH, Ferry GV (1972) Cosmic dust in the mesosphere. Space Res 12:369–380

Farlow NH, Ferry GV, Blanchard MB (1970) Examination of surfaces exposed to a noctilucent cloud, August 1, 1968. J Geophys Res 75:6736–6750

Fast NP (1972) Katalog pojavlenij serebristych oblakov po mirovym dannym. Tomsk Univ, Tomsk (in Russian)

Fast NP, Fast WH (1981) Letter dated 1st December 1981 to M. Gadsden

Fechtig H, Feuerstein M, Gerloff U, Weihrauch HJ (1969) Experimente zur Sammlung von kosmischen Staub mit Raketen. Vortrag AG "Extraterr Phys"

Fechtig H, Rauser P (1970) Kombinierte Nachweis- und Sammelexperimente für kosmische Staubteilchen. Vortrag AG "Extraterr Phys"

Fechtig H, Feuerstein M (1970) Particle collection results from a rocket flight on August 1, 1968. J Geophys Res 75:6751–6757

Fechtig H, Feuerstein M, Rauser P (1971) A simultaneous collection and detection experiment for cosmic dust. Space Res XI:335–346

Fedynski VV (1957) On the problem of noctilucent cloud wave movements. Astron Circ 181:25–27 (in Russian)

Ferguson EE (1971) D-region chemistry. Rev Geophys Space Phys 9:997–1008

Ferguson EE, Fehsenfeld FC (1968) Some aspects of the metal ion density of the earth's atmosphere. J Geophys Res 73:6215–6223

Fessenkov N (1947) Meteoric particles in the interplanetary space. Akad Nauk, Moscow Leningrad (in Russian)

Fiocco G (1967) The observation of noctilucent clouds by optical radar. Intern Symp Tallinn 1966, Moscow pp 95–96

Fiocco G, Grams G (1966) Observation of the upper atmosphere by optical radar in Alaska and Sweden during the summer 1964. Tellus 18:34–38

Fiocco G, Grams G (1969) Optical radar observations of mesospheric aerosol in Norway during the summer 1966. J Geophys Res 74:2453–2458

Fiocco G, Grams G (1971) On the origin of noctilucent clouds, extraterrestrial dust and trapped water molecules. J Atmos Terr Phys 33:815–824

Fiocco G, Smullin LD (1963) Detection of scattering layers in the upper atmosphere (60–140 km) by optical radar. Nature (London) 199

Fiocco G, Grams G, Visconti G (1975) Equilibrium temperatures of small particles in the Earth's upper atmosphere (50-110 km). J Atmos Terr Phys 37:1327-1337

Fogle B (1964) Geophys Inst Rep UAG R-158, Univ Alaska

Fogle B (1965) Noctilucent clouds over Punta Arena, Chile. Nature (London) 207:66

Fogle B (1966) Noctilucent clouds. UAG Rep 177, Univ Alaska

Fogle B, Haurwitz B (1966a) Noctilucent clouds. Space Sci Rev 6:279-340

Fogle B, Haurwitz B (1966b) Bonner Meteorol Abh 17:263

Fogle B, Haurwitz B (1969) Wave forms in noctilucent clouds. Deep-Sea Res 16:85-95

Fogle B, Rees MH (1972) Spectral measurements of noctilucent clouds. J Geophys Res 77:720

Francis RJ, Bennett JM, Seedsman DL (1969) Noctilucent clouds observed in Antarctica. Nature (London) 211:398

Gadsden M (1975) Observations of the colour and polarization on noctilucent clouds. Ann Geophys 31:507-516

Gadsden M (1978) The size of particles in noctilucent clouds: implications for mesospheric water vapor. J Geophys Res 83:1155-1156

Gadsden M (1980) A meteoric nightglow? Mon Not R Astron Soc 192:581-594

Gadsden M (1981b) The silver-blue cloudlets again: nucleation and growth of ice in the mesosphere. Planet Space Sci 29:1079-1087

Gadsden M (1982) Noctilucent clouds. Space Sci Rev 33:279-334

Gadsden M (1983a) A note on the orientation and size of noctilucent clouds particles. Tellus 35B:73-75

Gadsden M (1983b) The orientation of scatters in noctilucent clouds. Geophys Res Lett 10:889-891

Gadsden M (1985a) The earliest observations of noctilucent clouds. In: Schröder W (ed) Historical events and people in geosciences. Lang, Bern Frankfurt 1985, pp 131-139

Gadsden M (1985b) Observations of noctilucent clouds from NW-Europe. Ann Geophys 3:119-126

Gadsden M (1986a) Pedro Nuñez and the first printed treatise on twilight observations. Mitt Arbeitskreis Geschichte Geophys, Deutsche Geophys Ges 5:1-27

Gadsden M (1986b) Noctilucent clouds. Observatory 106:61

Gadsden M (1986c) Noctilucent clouds. QJR Astron Soc 27:351-366

Gadsden M (1986d) Noctilucent clouds in northwest Europe. Collect Works Int Workshop Noctilucent Clouds. Valgus, Tallinn pp 5-17

Gadsden M, Schröder W (1989) The nature of noctilucent clouds. Gerlands Beitr Geophys 98 (in press)

Gadsden M, Rothwell P, Taylor MJ (1979) Detection of circularly light from noctilucent clouds. Nature (London) 278:628-629

Gärtner V, Memmesheimer M (1984) Computation of the zonally-averaged circulation driven by heating due to radiation and turbulence J Atmos Terr Phys 46:755-765

Gärtner V, Memmesheimer M, Blum PW (1986) Seasonal transitions and possible polar mesospheric cloud regions calculated by a zonally averaged model of the middle atmosphere. J Atmos Terr Phy 48:1185-1196

Ganguly S (1984) Identification of mesospheric heavy ion ledge. J Atmos Terr Phys 46:99-103

Garcia RR, Solomon S (1985) The effects of breaking gravity waves on the dynamics and chemical composition of the mesosphere and lower thermosphere. J Geophys Res 90:3850-3868

Gavine DM (1984) Noctilucent clouds over western Europe during 1983. Meteorol Mag 113:272-277

Gavine DM (1985) Noctilucent clouds over western Europe during 1984. Meteorol Mag 114:349-356

Gavine DM (1986) Noctilucent clouds over western Europe 1985. Meteorol Mag 115:361-370

Geller MA (1979) Dynamics of the middle atmosphere. J Atmos Terr Phys 41:683-705

Gibbins CJ, Schwartz PR, Thacker DL, Bevilacqua RM, Hulburt EO (1982) The variability of mesospheric water vapor. Geophys Res Lett 9:131-134

Glöde P (1966) Ergebnisse der Radarmeteorbeobachtungen des Quandrantidenstroms. Kleinheub Ber 11:157

Götzelmann A (1985) Vorbereitung und Auswertung massenspektrometrischer Messungen von positiven Ionen in der polaren mittleren Atmosphäre im Winter. Dipl-Arbeit MPI Heidelberg

Goldberg RA, Witt G (1977) Ion composition in a noctilucent cloud region. J Geophys Res 82:2619–2627

Gosshard EE, Munk W (1954) On gravity waves in the atmosphere. J Meteorol 11:259–269

Grahn S, Witt G (1971) Rep AP-4 Inst Meteorol Univ Stockholm

Greenhow JS, Neufeld EL (1961) Winds in the upper atmosphere. QJR Meteorol Soc 87:472–489

Gregory JB (1958) The mesopause region of the ionosphere. Nature (London) 181:753–754

Griffiths L, Shanklin L (1987) An observation of noctilucent clouds in the Antarctic winter. Weather 42:391

Grishin NI (1955) On the structure of noctilucent clouds. Meteorol Gidrol 1:23–28 (in Russian)

Grishin NI (1956) On the structure of noctilucent clouds. Bjull VAGO 16:3–6 (in Russian)

Grishin NI (1961) Wave motions and meteorological conditions for the appearance of noctilucent clouds. Ann IGY 11:20–22

Grishin NI (1967a) Dynamical morphology of noctilucent clouds. In: Noctilucent clouds, Akad Nauk, Moscow pp 193–199

Grishin NI (1967b) On the possible composition of noctilucent clouds. Astron Vestn 1:89–96 (in Russian)

Grishin NI (1970) Observations of noctilucent clouds on July 3–4, 1967. Fiz mezsosfern (serebrist) oblakov. Riga 1970 pp 165–174 (in Russian)

Grishin NI, Kurilova JV (1973) Morphology and dynamics of noctilucent clouds. Noctilucent clouds. Optical properties, Akad Nauk, Tallinn pp 56

Grjebine T (1967) Abundances of cosmic dust. In: Runscorn SC (ed) Mantles of the earth and terrestrial planetes. Academic, London pp 63–73

Gromova TD (1963) Catalogue of observations of noctilucent clouds on the territory of our country for 1885–1956. Issl serebrist oblakov 6:64–83 (in Russian)

Grossmann KU, Frings WG, Offermann D, Andre L, Kopp E, Krankowsky D (1985) Concentration of H_2 and NO in the mesosphere and the lower thermosphere at high latitudes. J Atmos Terr Phys 47:291–300

Groves GV (1968) Comparisons of new results with CIRA 1965 with emphasis on the IQSY data a review for the region 40 to 100 km. Space Res VIII:741–760

Groves GV (1971) AFCRL Environm Res Pap 368

Guthnick M (1961) How dry is the sky? J Geophys Res 66:2867–2871

Hallert B (1960) Photogrammetry basic principles and general survey. McGraw-Hill, New York

Hallgren DS, Hemenway CL, Mohnen VA, Tackett DC (1973a) Preliminary results from the noctilucent cloud sampling from Kiruna in 1970. Space Res XIII:1099–1104

Hallgren DS, Hemenway CL, Schmalberger DC (1973b) Noctilucent cloud sampling by a multi-experiment payload. Space Res XIII:1105–1122

Hamilton RA (1964) Observation of noctilucent clouds near midwinter. Meteorol Mag 93:201–202

Hanson AM (1963) Noctilucent clouds at 76°3 north. Weather 18:142–144

Hanson AM (1965) Noctilucent clouds over the arctic in November. J Geophys Res 70:4717–4718

Hapgood MA, Taylor MJ (1982) Analysis for airglow image data. Ann Geophys 38:805–815

Harrison AW (1973) Spectrophotometric measurements of noctilucent clouds. Can J Phys 51:573–577

Hartman GK, Kunzi KF, Lobsiger E (1983) Mesospheric water vapour measured with an airborne sensor. In: Wave propagation and remote sensing. ESA, Paris, pp 269–271

Hass H, Ebel A (1986) Space and time scales of large scale variations in the upper stratosphere and mesosphere as deduced from PMR of Nimbus 6. J Atmos Terr Phys 48:1073–1083

Haurwitz B (1961) Wave formations in noctilucent clouds. Planet Space Sci 5:92–98

Haurwitz B (1964) Comments on wave forms in noctilucent clouds. UAG-Rep 160, Univ of Alaska

Heard WC (1968) Mesospheric-stratospheric coupling. Nature (London) 204:1038–1040

Heintzenberg J, Witt G, Kinmark I (1978) Optical characteristics of noctilucent clouds: measurements and interpretation. Proc 6th Ann Meet Upper Atmos Stud Opt Methods pp 78–84

Hemenway CL, Soberman RK (1962a) Studies of micrometeorites obtained from a recoverable sounding rocket. Astron J 67:256

Hemenway CL, Witt G (1963) Particle sampling from noctilucent clouds. Nature (London) 199:269–270

Hemenway CL, Fullam EF, Skrivanek RA, Soberman RK, Witt G (1964) Electron microscope studies of noctilucent clouds particles. Tellus 16:96–102

Hemenway CL, Soberman RK, Witt G (1964) Sampling of noctilucent cloud particles. Tellus 16:84–88

Hemenway CL, Hallgren DS, Kerridge JF (1968) Results from the Gemini S-10 and S-12 micrometeorites experiments. Space Res VIII:521–535

Hersé M (1989) Bright nights. In: Schröder W (ed) Past, present and future trends in geophysical research. IAGA Interd. Comm History Bremen- Rönnebeck 1988 pp 41–64

Hesstvedt E (1961) Note on the nature of noctilucent clouds. J Geop Res 66:1985–1987

Hesstvedt E (1962) On the possibility of ice clouds formation at the mesopause. Tellus 14:290–296

Hesstvedt E (1964) On the water content of the high atmosphere. Geofys Publ 25:1–18

Hesstvedt E (1969a) Nucleation and growth of noctilucent clouds particles. Space Res IX:170–174

Hesstvedt E (1969b) Noctilucent clouds observations and deduction: water vapor in the stratosphere and mesosphere. Ann IQSY 1969:23

Heymsfield AJ (1986) Ice particles observed in a cirrus form cloud at -83° and implication for polar stratospheric heights. J Atmos Sci 43:851–853

Hines CO (1959) An interpretation of certain ionospheric motions in terms of atmospheric waves. J Geophys Res 64:2210–2211

Hines CO (1960) Internal atmospheric gravity waves at ionospheric heights. Can J Phys 38:1441–1481

Hines CO (1968a) Some consequences of gravity waves critical layers in the upper atmosphere. J Atmos Terr Phys 30:845–856

Hines CO (1968b) A possible source of waves in noctilucent clouds. J Atmos Sci 25:937–942

Hines CO, Reddy CA (1967) On the propagation of atmospheric gravity waves through regions of wind shear. J Geophys Res 72:1015–1034

Hobbs PV (1974) Ice physics, 441, 462, Clarendon, Oxford

Hoffmeister C (1951) Spezifische Leuchtvorgänge im Bereich der mittleren Ionosphäre. Erg ex Naturwiss 24:1–51

Hoffmeister C (1961) Nature and origin of noctilucent clouds. Ann IGY 11:13–15

Holton JR (1972) An introduction to dynamical meteorology. Academic Press, London New York

Holton JR (1975) The dynamics meteorology of the stratosphere and mesosphere. Meteorol Monogr AMS 37:218

Houghton JT (1977) The physics of the atmosphere. Cambridge Univ Press, Cambridge

Houghton JT (1978) The stratosphere and mesosphere. QJR Meteorol Soc 104:1–29

Hulst van de G (1957) Light scattering by small particles. Willey and Sun, New York

Hummel JR (1975) Satellite observation of the mesospheric scattering layer and implied climatic consequences, Ionos Res Lab Rep PSU-IRL-IR-52, Pennsylvania State Univ

Hummel JR (1977) Contribution to polar albedo from a mesospheric aerosol layer. J Geophys Res 82:1893–1900

Hummel JR, Olivero JJ (1976) Satellite observations of the mesospheric scattering layer and implied climatic consequences. J Geophys Res 81:3177–3178

Hunt BG (1971) Cluster ions and nitric oxide in the D-region. J Atmos Terr Phys 33:929–942

Hunt BG (1987) Excitation and propagation of middle atmosphere internal gravity waves in a general circulation model. J Atmos Terr Phys 49:339–351

Hunten DM, Turco RP, Toon OB (1980) Smoke and dust particles of meteoric origin in the mesosphere and stratosphere. J Atmos Sci 37:1342–1357

Ivanovsky AI (1966) On the relation of atmospheric stability to low-frequency elastic oscillation with the wave structure of noctilucent clouds. Proc Noctilucent Clouds. Tallinn, pp 216–221

Iwasakawa Y (1982) Formation of noctilucent cloud particles and the temperature distribution at the polar mesopause. Jpn Natl Pol Res Inst Mem Spec Issue 22:247–253

Izakov MN, Kokin GA, Perov SB (1967) Metodika izmerenlja davlenija i temperaturyi pri meteorologi'eskom raketnom zondipovanii. Meteorol Gidr 12:70–86 (in Russian)

Jakobs H, Bischof M, Ebel A, Speth R (1986) Simultation of gravity wave effects under solstice condition using a 3-D circulation model of the middle atmosphere. J Atmos Terr Phys 48:1203–1223

Jensen E, Thomas GE (1988) A growth sedimentation model of polar mesospheric clouds: comparison with SME measurements. J Geoph Res 93:2461–473

Johannessen A, Krankowsky D, Arnold F (1972) Detection of water cluster ions at the high latitude summer mesopause. Nature (London) 235:215–217

Johnson FS (1975) Transport processes in the upper atmosphere. J Atmos Terr Phys 32:1658–1662

Johannessen A, Thrane EV (1974) Rocket study of the high-latitude summer mesosphere. Indian J Radio Space Phys 3:128–132

Joseph JH (1967) Detection of noctilucent clouds in the twilight layer from satellites. J Geophys Res 72:4020–4025

Juskeseleiva L (1963) Condensation hypothesis of noctilucent cloud formation. Bulg Izv Geofys Bl'gn AN 4:25–33

Kaiser TR (1961) The incidence of interplanetary dust. Ann Geophys 17:50–59

Kaiser TR (1962) Meteors and the abundance of interplanetary matter. Space Sci Rev 1:544–575

Kallmann-Bijl KH (1963) Variation of atmospheric properties with time and solar activity. Space Res III:78–88

Kashtanov AF, Novikov BM (1981) Role of Hydrogen (H_2O_2) in the formation of mesosphere clouds. Meteorol Gidr 7:105–107 (in Russian)

Kato S (1980) Dynamics of the upper atmosphere. Reidel, Dordrecht

Keegan TJ (1961) Winds and circulations in the mesosphere. J Am Rock Soc 31:1060–1066

Kellogg WW (1961) Chemical heating above the polar mesosphere in winter. J Meteorol 18:373–381

Kellogg WW, Schilling GF (1951) A proposed model of the circulation in the upper atmosphere. J Meteorol 8:220–230

Kerker M (1969) The scattering of light and other electromagnetic radiation. Academic Press, London, New York

Khantodze AG (1976) Wellenbewegungen in der Ionosphäre. Phys Solarterr 1:41–59

Kinnmarck I (1979) Numerical study of particle growth and water vapor concentration in noctilucent clouds. Rep A-18, Dep Meteorol, Univ Stockholm

Kochanski A (1963) Circulation and temperature at 70 to 100 kilometer height. J Geophys Res 68:213–226

Kohl G (1967) Depolarisation des Zenitdämmerungslichts, leuchtende Nachtwolken und hohes Ozon. Gerlands Beitr Geophys 76:353–363

Kohl G (1972) Der qualitative Nachweis von Schichten kosmischer Staubpartikeln in der unteren Thermosphäre mit Hilfe der dämmerungs-optischen Methode. Bull Abustuman Astrophys Obs 41:27–48

Konasenok VN, Sved GM (1968) On the physical condition near the mesopause. Fiz atmosf i okeana 4:490–498 (in Russian)

Kopp E, Herrmann U (1984) Ion composition in the lower ionosphere. Ann Geophys 2:83–94

Kopp E, Bertin F, Björn LG, Dickinson PHG, Philbrick G, Witt G (1982) The "Camp" campaign. ESA SP-229

Kopp E, Eberhardt P, Herrmann U, Björn LG (1985) Positive ion composition of the high-latitude summer D-region with noctilucent clouds. J Geophys Res 90:13041–13053

Kosibowa S (1972) Historia vadan oblokow mezsosfercznych. Przeg Geofiz 16:341–352 (in Polish)

Kosibowa S, Pyka JL (1973) Mesospheric clouds visible in Poland during 1971. Acta Geophys Pol 21:169–178

Kosibowa S, Speil J, Pyka JL (1975) Mesospheric clouds observed in Poland in 1973. Acta Geophys Pol 23:91–103

Kosibowa S, Speil J, Pyka JL (1976) Mesospheric clouds observed in Poland during 1974. Acta Geophys Pol 24:181–193

Kosibowa S, Pyka JL, Speil J (1978) Mesospheric clouds observed in Poland during 1976. Acta Geophys Pol 26:63–81

Krassovsky VI (1980) Nitric oxide, water steam, noctilucent cloud emission and radio wave absorption near the mesopause. Geomagn Aeronomy 20:657

Kropotkina EP, Shefov NN (1975) Vliyane lunnikh prilivov na veroyatnost poyaleniya serebristych oblakov. Izv Atmosf Okean Fiz 11:1184–1186 (in Russian)

Kuhnke F (1976) Leuchtende Nachtwolken als Indikator für Wellenerscheinungen in der Mesopause. Dipl-Arbeit, Univ Braunschweig

Kurilova JV (1962) Aero-synoptical condition of the formation of noctilucent clouds. Trudy sovesc serebrist oblak Tallinn, pp 131–150 (in Russian)

Kurilova JV (1966) On meteorological conditions of the occurrence of noctilucent clouds. Serebrist Oblakam 12:101–105

Latimer P, Barber P (1978) Scattering by ellipsoids of revolution: a comparison of theoretical methods. J Colloid Interface Sci 63:310–316

Lauter EA (1962) Ein Beitrag zum Nachweis der Auswirkungen des Sonnenfleckenzyklus in der tieferen Ionosphäre. Gerlands Beitr Geophys 71:58–61

Lauter EA (1974) Mesospheric properties as seen from D-region electron density behavior. Z Meteorol 24:65–79

Lauter EA, Sprenger K (1967) Erscheinungen in der D-Region im Zusammenhang mit Strukturänderungen des Neutralgases der Strato- und Mesosphäre. Kleinheub Ber 12:297–301

Lauter EA, Entzian G, Cossart G von, Sprenger K, Greisinger KM (1977) Synoptische Erschließung von Prozessen der winterlichen Mesopausenregion durch bodengebundene Beobachtungsverfahren. Z Meteorol 27:75–84

Lauter EA, Entzian G (1982) Solarterr 19:118

Lazarev AI, Leonov AA (1973) Optical observations from the Voskhod-2 manned spacecraft. Opt Tech 40:410–413

Leovy BC (1961) Simple models of thermally driven mesospheric circulation. J Atmos Sci 21:327–341

Leovy CB (1966) Photochemical destabilization of gravity waves the mesopause. J Atmos Sci 23:223–232

Lilley EA, Radford HE, Litvak MM, Gottleib CB, Rosenthal SK (1971) Mesospheric water vapour measured from ground-based microwave observations. J Geophys Res 82:472–478

Lindzen RS (1970) Internal gravity waves in atmospheres with realistic dissipation and temperature. Geophys Fluid Dyn 1:303 and 2:31

Linscott J, Hemenway LJ, Witt G (1964) Calcium film indicator of moisture associated with noctilucent clouds particles. Tellus 16:110–113

Liu SC, Donahue TM (1974) The aeronomy of hydrogen in the atmosphere of the earth. J Atmos Sci 31:1118–1136

Llyod KH, Low CH, Vincent RA (1973) Turbulence, billows and gravity waves in a high-shear region of the upper atmosphere. Planet Space Sci 21:653–661

Lowan AN (1955) On the cooling of the upper atmosphere after sunset. J Geophys Res 67:421–429

Ludlam FH (1957) Noctilucent clouds. Tellus 9:341–364

Martin H (1966) Die Tunguska-Katastrope in geophysikalischer Hinsicht. Sterne 42:45

Martynkevic GM (1971) O prirode mezosfernych oblakov. Meteorol Gidr 12:44 (in Russian)

Martynkevic GM (1973) Atomic hydrogen and noctilucent clouds. Trudy sovesc serebrist oblakam Tallinn, p 43 (in Russian)

McDonald JE (1964) On a criterion governing the mode of cloud formation in planetary atmospheres. J Atmos Sci 21:76–83

McIntosh DH, Hallissey M (1974) Noctilucent clouds over western Europe during 1973. Meteorol Mag 103:157–160

McIntosh DH, Hallissey M (1975) Noctilucent clouds over western Europe and the Atlantic during 1975. Meteorol Mag 105:18

McIntosh DH, Hallissey M (1976) Noctilucent clouds over western Europe and the Atlantic during 1976. Meteorol Mag 105:187–191

McIntosh DH, Hallissey M (1977) Noctilucent clouds over western Europe and the Atlantic during 1976. Meteorol Mag 106:181–184

McIntosh DH, Hallissey (1978) Noctilucent clouds over western Europe. Meteorol Mag 107:182–187

McIntosh DH, Hallissey M (1979) Noctilucent clouds over western Europe Meteorol Mag 108:185–189

McIntosh DH, Hallissey M (1980) Noctilucent clouds over western Europe. Meteorol Mag 109:182–184

McIntosh DH, Hallissey M (1981) Noctilucent clouds over western Europe. Meteorol Mag 110:109–112

McIntosh DH, Hallissey M (1982) Noctilucent clouds over western Europe. Meteorol Mag 111:122–125

McKay Ch (1982) A photochemical-thermal model of hydrogen and oxygen in the summer mesosphere and implications from noctilucent clouds formation. Thesis, Univ Colorado, Boulder

McKay Ch (1985) Noctilucent clouds formation and the effects of water vapor variability on temperatures in the middle atmosphere. Planet Space Sci 33:761–771

McLone RR (1966) On the form of noctilucent clouds. PAGEOPH 67:233–238

Megrelishvili TG, Chvostikov IA (1970) Atomic hydrogen, oxygen and H_2O molecules in the upper atmosphere and some problems of geophysics and physics near space. Fizika mezosfern (serebrist) oblakov, Riga pp 9–24 (in Russian)

Meier L (1955) Bearbeitung einer Erscheinung von leuchtenden Nachtwolken. Sterne 31:230–234

Meinel AB, Middlehurst B, Whitaker E (1963) Low latitudes noctilucent clouds. Science 141:1176–1178

Memmesheimer M, Blum PW (1988) Seasonal and latitudinal changes in atmospheric condition favoring the formation of polar mesopheric clouds. Phys scripta 37:178–184

Memmesheimer M, Gärtner V, Thomas EG, McKay CP (1985) The role of horizontally-averaged nonlinear heat transport in zonally-averaged models of the middle atmosphere. Ann Geophys 3:187–194

Millmann PM (1959) Visual and photographic observations of meteors and noctilucent clouds. J Geophys Res 64:2122–2128

Minnaert M (1954) The nature of light and color in the open air. Dover Publ, Dover

Moreels G, Hersé M (1977) Photographic evidence of waves around the 85 km level. Planet Space Sci 25:266–273

Morozov VM (1956) Problem of the origin of noctilucent clouds. Izv Akad Nauk SSSR Ser Geofiz 7:865–869 (in Russian)

Murgatroyd RJ (1957) Winds and temperatures between 20 and 100 km – a review. QJR Meteorol Soc 83:417–458

Murgatroyd RJ, Singleton F (1961) Possible meridional circulation in the stratosphere and mesosphere. QJR Meteorol Soc 87:125–135

Narcisi RS, Bailey AD, Della Lucca L, Sherman C, Thomas DM (1971) Mass spectrometric measurements of negative ions in the D- and E-regions. J Atmos Terr Phys 33:1147–1159

Nastrom GD, Balsly BB, Carter DA (1982) Mean meridional winds in the mid- and high latitude summer mesosphere. Geophys Res Lett 9:139–142

Nestorov G (1972) Lower ionosphere at medium latitudes along geomagnetic disturbances. Bull Geophys Inst BAW 18:19

Nicolet M (1984) On the photodissociation of water vapor in the mesosphere. Planet Space Sci 32:871–880

NOAA (1976) US Standard Atmosphere 1976. NOAA-S/T76-1562, US Gov Printing Office, Washington DC

Nordberg W, Stroud WC (1961) Seasonal, latitudinal and diurnal variations in the upper atmosphere. NASA Tech Note D-703

Nordberg W, Smith WS (1963) Rocket measurements of the structure of the upper stratosphere and mesosphere. Meteorol Abh FU Berlin XXXVI:391

Nordberg W, Katchen L, Theon J, Smith WS (1965) Rocket observations of the structure and dynamics of the mesosphere during quiet sun period. NASA Publ X-651-65-154

Nordberg W, Katchen L, Theon J, Smith WS (1965b) Rocket observations of the structure of the mesosphere. J Atmos Sci 22:611–622

Novozilov NJ (1962) On the role of wave motions in noctilucent clouds in the study of mesopause structure. Trudy sovesc po serebrist oblakam Tallinn, pp 126–130 (in Russian)

Oleak H (1956/57) Das Verhalten von Mikrometeoriten in der Erdatmosphäre. Wiss Z Univ Jena Math-Naturwiss Reihe 6:133–143
Oleak H (1961) Die Bewegung von Mikrometeoriten in der Erdatmosphäre. Sterne 37:67–72
Olivero JJ, Bevilacqua RM (1979) Physical properties affecting the existence of small ice particles in the mesosphere. Space Res 19:165–168
Olivero JJ, Thomas GE (1981) Climatology of polar mesospheric clouds. J Atmos Sci 43:1263–1274
Olivero JJ, Gibbins JJ, Bevilacqua RM, Schwartz PR, Thacker DL (1981) Intern Assoc Geomag Aeron 5th Sci Ass Edinburgh
Olivero JJ, Tsou JJ, Croskey CL, Hale LC, Joiner RG (1986) Solar absorption microwave measurements of upper atmospheric water vapor. Geophys Res Lett 13:197–200
Ordt Auff'm N, Brodhun D (1974) Zur Deutung der Wellenstrukturen auf leuchtenden Nachtwolken. Z Meteorol 24:291–298
Orr C, Keng EYA (1964) Photophoretic effects in the stratosphere. J Atmos Sci 21:475–478
Packer DM, Packer IG (1977) Appl Opt 16:983–992
Parthasarathy R (1976) Mesopause dust as a sink for ionization J Geophys Res 81:2392–2396
Paton J (1950) Aurora and luminous night clouds. Proc Phys Soc B 63:1039–1040
Paton J (1954) Direct evidence of vertical motion in the atmosphere at a height of about 80 km provided by photographs of noctilucent clouds. Proc Toronto Meteorol Conf 1953:31–33
Paton J (1961) Noctilucent clouds. Ann IGY 11:4–6
Paton J (1964) Noctilucent clouds. Meteorol Mag 93:161–179
Paton J (1965) Noctilucent clouds over western Europe during 1964. Meteorol Mag 94:180–184
Paton J (1966) Noctilucent clouds over western Europe during 1965. Meteorol Mag 95:174–176
Paton J (1967) Noctilucent clouds over western Europe during 1966. Meteorol Mag 96:187–190
Paton J (1968) Noctilucent clouds over western Europe during 1967. Meteorol Mag 97:174–176
Paton J (1969) Noctilucent clouds over western Europe during 1968. Meteorol Mag 98:219–222
Paton J (1970) Noctilucent clouds over western Europe during 1969. Meteorol Mag 99:184–186
Paton J (1971) Noctilucent clouds over western Europe during 1970. Meteorol Mag 100:179–182
Pavlova TD (1960) Apparent frequency of the appearance of noctilucent clouds based on observations at the stations of the hydrometeorological service for 1957 and 1958. Izd Astron obser issl serebrist oblakov 1:3–58 (in Russian)
Pavlova TD (1962) A study of longitudinal distribution of noctilucent clouds. Trudy sovesc po serebrist oblakam Tallinn, pp 119–125 (in Russian)
Philbrick C, Fiare AC, Fryklund DH (1984) The state experiment – mesospheric dynamics. Adv Space Res 4:153–156
Pollermann R (1972) Massenspektrometrische Messungen positiver Ionen in der unteren Ionosphäre bei Anwesenheit leuchtender Nachtwolken. Thesis, Univ Heidelberg
Quiroz RS (1964) On the origin and climatology of noctilucent clouds. Tech Rep 181 AWS (MATS) USAG
Quiroz RS (1961) Seasonal and latitudinal variations of air density in the mesosphere (30 to 80 kilometers). J Geophys Res 66:2129–2139
Radford HE, Litvak MM, Gottlieb CA, Gottlieb EW, Rosenthal SK, Lilley AE (1977) Mesospheric water vapor measured from ground based microwave observations. J Geophys Res 82:472–478
Rajchl J (1968) Fireballs and noctilucent clouds. BAC 37:305–311
Rauser P, Fechtig H (1972) Combined dust collection and detection experiment during a noctilucent cloud display above Kiruna, Sweden. Space Res XII:391–402
Rauser P, Fechtig H (1973) Dust measurements in the upper atmosphere during and in the absence of noctilucent cloud display. Space Res XIII:1127–1133
Reid GC (1975) Ice clouds at the summer polar mesopause. J Atmos Sci 32:523–535
Reid GC (1977) The production of water cluster positive ions in the quiet daytime D-region. Planet Space Sci 25:275
Reid GC, Solomon S (1986) On the existence of an extraterrestrial source of water vapor in the middle atmosphere. Geophys Res Lett 13:1129–1132
Reiter ER, Haurwitz B (1974) Internal gravity waves in the atmosphere Arch Meteorol Geophys Bioklim A 23:101–114

Roddy JW (1986) The physics of noctilucent clouds formation. Collec Pap Int-Workshop Tallinn, pp 33-58

Rössler F (1972) Aerosol layers in the atmosphere. Space Res XII:423-431

Roper RH (1966) Dissipation of wind energy in the height of 80-140 km. J Geophys Res 71:4427-4428

Rosenberg NW, Edwards HP (1964) Winds and shears in the lower ionosphere. 5th Conf Appl Meteorol, March 2-6, 1964

Rosenfeld Shk (1984) Drift motions of aerosols particles in the atmospheric gravity waves. Izv Akad SSR Fiz Atmos Okean 20:33-39 (in Russian)

Rosenfeld Shk (1986) Relationship between observations in noctilucent clouds and internal gravity waves. In: Collect Works Int Workshop Noctilucent clouds. Tallinn, pp 173-192

Rosinski J, Pierrard JM (1964) Condensation products of meteor vapors and their connection with noctilucent clouds and rainfall. J Atmos Terr Phys 26:51-56

Rozenberg GV (1963) Twilight Fitzmatgiz, Moscow (in Russian)

Scherhag R (1969) Uber Luftdruck-, Temperatur- und Windschwankungen in der Stratosphäre. Abh Akad Wiss Lit Mainz Math-naturwiss Kl 15:1439-1528

Schmitz G (1981) Zum Transport chemisch aktiver Bestandteile durch transiente planetare Wellen in Stratosphäre und Mesosphäre. Z Meteorol 31:300-305

Schröder W (1962) Observations of upper atmospheric light phenomena during 1960 and 1961. Gerlands Beitr Geophys 71:239-241

Schröder W (1964) On noctilucent clouds. Gerlands Beitr Geophys 73:157-160

Schröder W (1965) Simultaneous sightings of aurora and noctilucent clouds. Gerlands Beitr Geophys 74:471-473

Schröder W (1966a) On the seasonal frequency of noctilucent clouds. Meteorol Rundsch 19:91-93

Schröder W (1966b) On the diurnal frequency of noctilucent clouds. Meteorol Rundsch 19:26-27

Schröder W (1966c) Results and problems of noctilucent clouds investigation. Gerlands Beitr Geophys 75:45-55

Schröder W (1967a) On the 'latitudinal shift' of noctilucent clouds. Meteorol Rundsch 20:54-55

Schröder W (1967b) The noctilucent cloud display of 1967, July 4-5. Meteorol Rundsch 20:176-177

Schröder W (1967c) Studies on noctilucent clouds during the years 1957-1966 in Germany. Gerlands Beitr Geophys 76:139-144

Schröder W (1967d) Widespread occurrence of noctilucent clouds. Z Meteorol 19:109-111

Schröder W (1967e) Results of the study of noctilucent clouds over Germany during 1885-1965. J Geophys Res 72:1971-1973

Schröder W (1968a) Spring transition of the mesosphere and frequency of noctilucent clouds. Gerlands Beitr Geophys 77:191-194

Schröder W (1968b) On the relation between mesospheric circulation and frequency of occurrence of noctilucent clouds. Gerlands Beitr Geophys 77:303-308

Schröder W (1968c) Analysis of the diurnal frequency of noctilucent clouds. Meteorol Rundsch 21:28-30

Schröder W (1968d) Mesospheric circulation and noctilucent clouds. Meteorol Rundsch 21:54-56

Schröder W (1969) On the kinematic of noctilucent clouds. Gerlands Beitr Geophys 79:229-235

Schröder W (1970) Aurora and noctilucent clouds. Gerlands Beitr Geophys 79:223-228

Schröder W (1971) Investigations of the transition periods in spring and autumn of the mesopause. Gerlands Beitr Geophys 80:65-74

Schröder W (1974) Some aspects of the noctilucent clouds and mesospheric circulation. Idöjárás 78:31-40

Schröder W (1975) Developmental phases of noctilucent cloud research. Akademie-Verlag, Berlin

Schröder W (1984) Das Phänomen des Polarlichts (Aurora Borealis). Wissenschaftl Buchges, Darmstadt

Schröder W (1985) Krakatoa 1885. Geowiss in unserer Zeit 1:155-159

Schröder W (1987) Note on the seasonal frequency of polar mesospheric clouds and noctilucent clouds. Prepr 1/1987

Schröder W (1988a) Polar mesospheric clouds and noctilucent clouds: a comparison. Preprint
    2/1988
Schröder W (1988b) An observation of noctilucent clouds in the Antarctic winter. Weather 43:380
Schulte P (1984) Massenspektrometrische Messung schwerer Ionen in der Mesopausenregion
    während des arktischen Sommers. Dipl-Arbeit, Univ Heidelberg
Scott AFD (1972) Mesospheric temperatures and winds during a stratospherir warming. Philos
    Trans R Soc London Ser A 271:547
Self S, Rampino MR (1981) The 1883 eruption of Krakatoa. Nature (London) 294:699–704
Sergin SJA, Sargin VJA (1967) Formative processes of noctilucent clouds. Meteorol Gidr 4:51–56
    (in Russian)
Shafrir U, Humi M (1967) Interplanetary particles and noctilucent clouds. J Atmos Sci 25:577–581
Sharonov VV (1963) Problems of noctilucent clouds climatology. Meteorol Issl 6:5–22 (in Russian)
Shefov NN (1967) Proc Int Symp Noct Clouds, Tallinn 1966, pp 187–189
Sheshamani R (1977) Geomagnetic activity effects and mesosphere temperature. Space Res XVII:
    141–146
Shurcliff WA (1962) Polarized light production and use. Oxford Univ Press, Oxford
Simmons DAR (1977) Weather 32:240–248
Simmons DAR, McIntosh DH (1983) An analysis of noctilucent clouds over western Europe during
    the period 1966 to 1982. Meteorol Mag 112:289–298
Sitarski M, Kerker M (1984) Monte Carlo simulation of phophoresis of submicron aerosol particles.
    J Atmos Sci 41:2250
Skrivanek RA, Soberman RK (1964) Simulation rings pattern observed with noctilucent cloud
    particles. Tellus 16:114–117
Smith D, Church MJ (1977) Ion recombination rates in the earth's atmosphere. Planet Space Sci
    25:433–439
Smith W, Katchen L, Sacher P, Swartz P, Theon J (1964) Temperature pressure, density and wind
    measurements with the rocket grenade experiment 1960–1963. NASA TR R-211
Soberman RK (1968) Extraterrestrial origin of noctilucent clouds. Paper Pres Cospar-meeting,
    Tokyo
Soberman RK, Chrest SA, Mannig L, Rey T, Ryan TG, Skrivanek RA, Wilhelm N (1964)
    Techniques for rocket sampling of noctilucent cloud particles. Tellus 16:89–95
Solomon S, Ferguson EE, Fehsenfeld DW, Crutzen PJ (1982) On the chemistry of H_2OH_2 and
    meteoric ions in the mesosphere and lower thermosphere. Planet Space Sci 30:1117–1126
Stroud WG, Nordberg WG, Bandeen WR, Bartman FL, Titus P (1960) Rocket grenade meas-
    urements of temperatures and winds in the mesosphere over Churchill, Canada. J Geophys Res
    65:2307–2323
Tarasova TM (1962) Polarization of the light flux of noctilucent clouds. Trudy sovesc serebr
    oblakam, Tallinn pp 55–67 (in Russian)
Taubenheim J (1975) Physik der Hochatmosphäre. In: Physik des Planeten Erde, hrsg v R
    Lauterbach, Enke, Stuttgart pp 80–116
Taubenheim J (1987) Aspekte der mittleren Atmosphäre im Problem der anthropogenen und
    exogenen Klimabeeinflussung. Veröff Forsch Ber Geo-Kosmoswiss 13:7–13
Taylor MJ (1981) A wide field lowlight level TV system to measure the state of polarization of light.
    J Phys E-Sci Instrum 14:865–869
Taylor MJ, Hapgood MA, Simmons DAR (1984) The effect of atmospheric screening of the visible
    border of noctilucent clouds. J Atmos Terr Phys 46:363–372
Theon JS, Nordberg GW, Katchen W (1966) Some observations on the thermal behaviour of the
    mesosphere. NASA Publ X-621-66-490
Theon JS, Nordberg W, Smith WS (1967) Temperature measurements in noctilucent clouds.
    Science 157:419–421
Theon JS, Smith WS (1969a) Seasonal transitions in the thermal structure of the mesosphere at high
    latitude. GSFC X-621-69-393
Theon JS, Smith WS, McGovern WE (1969b) Wind measurements in noctilucent clouds. Science
    164:715
Theon JS, Smith WS, McGovern WE (1970) Seasonal transitions in the thermal structure of the
    mesosphere at high latitude. J Atmos Sci 27:173–176

Thomas GE (1984) Solar mesosphere explorer measurements of polar mesospheric clouds (noctilucent clouds). J Atmos Terr Phys 46:819–824

Thomas GE, McKay CP (1985) On the mean particle size and water content of polar mesosphere clouds. Planet Space Sci 33:1209–1224

Thomas GE, McKay CP (1986) Mean particle sizes and water contents of polar mesospheric clouds. Coll Pap Workshop Nocticulent Clouds, Tallinn, pp 59–120

Thomas GE, Olivero JJ, Jensen EE, Schröder W, Toon OB (1989) Relation between increasing methane and the presence of ice clouds at the mesopause. Nature 338:490–492

Thomas GE, Olivero JJ (1986b) Heights of polar mesospheric clouds. Geophys Res Lett 13:1403–1406

Thomas L (1976) Mesospheric temperature and the formation of water cluster ions in the D-region. J Atmos Terr Phys 38:1345–1450

Thomas TF, Paulson JF (1979) On the photochemical stability of $H_3O + (H_2O)_n$. J Chem Phys 71:552–553

Tolstoy I (1967) Long-period gravity waves in the atmosphere. J Geophys Res 72:4605–4622

Toon O, Farlow NH (1981) Particles above the tropopause: measurements and models with stratospheric aerosols, meteoric debrism, nacreous clouds and noctilucent clouds. Annu Rev Earth Planet Sci 9:19

Tozer WF, Beeson DE (1974) Optical model of noctilucent clouds based on polarimetric measurements from two sounding rocket campaign. J Geophys Res 79:5607–5612

Trubinkov BN, Skuratova IS (1966) Cellular convection in the zone of noctilucent clouds. Proc Noctilucent clouds Int Symp, Tallinn pp 208–215

Turco RP, Toon OB, Whitten RC, Keese RG, Hollenbach D (1982) Noctilucent clouds: stimulation studies of their genesis, properties and global influences. Planet Space Sci 30:1147–1181

Twelves S (1957) Temperature and wind fields at the time of noctilucent clouds in Alaska, July 27–28, 1957. Mon Weather Rev 85:281

Vasilyev NV, Fast NP (1972) Anomal'nye opticeskie javöenija svazannyc s padeniem Tungosskogo meteorite. Gerlands Beitr Geophys 81:433–438 (in Russian)

Vasilyev OB (1962) Results of the absolute photometry and polarimetry of noctilucent clouds. Trudy sovesc serebr oblakam Tallinn pp 14–28 (in Russian)

Vasilyev OB (1967) Astrofiziceskie issledovanija serebristych oblakov. Astrofiz soobs Inf 4:1–86 (in Russian)

Vasilyev OB (1970) Fizika mezosfern (serebrist) oblakov, Riga p 121 (in Russian)

Veselov DP, Popov OI, Semenova VI, Seleznev GI, Fedorova Ye O (1976) Spectral brightness of noctilucent clouds in the visible and near infrared. Izv Atmos Ocean Phy 12:1097–1099

Vestine EH, Deirmendijan (1961) Some remarks on the nature and origin of noctilucent cloud particles. Ann IGY 11:6–13

Vincent RA, Ball S (1977) Tidal and gravity waves in the mesosphere a* mid and low latitudes. J Atmos Terr Phys 39:965–970

Vincent RA, Ball S (1984) MF/HF radar measurements of the dynamics of the mesopause regional review. J Atmos Terr Phys 46:961–974

Von Cossart GV, Entzian G (1976) Ein Modell der Mesopausenregion zur Interpretation indirekter Phasenmessungen und zur Abschätzung von Ionosphären- und Neutralgasparametern. Z Meteorol 26:219–230

Von Cossart GV, Taubenheim J (1987) Solar cycle and long-period variations of mesospheric temperatures. J Atmos Terr Phys 49:303–307

Wait JR (1955) Scattering of a plane wave from a circular dielectric cylinder at oblique incidence. Can J Phys 33:189–195

Webb W (1965) Morphology of noctilucent clouds. J Geophys Res 70:4463–4475

Willmann Ch (1962) On the polarization of light from noctilucent clouds. Trudy sovesc serbr oblakam Tallinn, pp 29–54 (in Russian)

Willmann ChI, Klimuk PI, Koksharov PI, Sevastyanov VI, Sergeyevich VN, Eerme KA (1977) Opt Issl Izpuch Atmos Pol Siyanii i serbr oblakov borta orbit nauch stach 'salyut-4', Tartu, pp 53–66 (in Russian)

Willmann Ch, Grechko GM, Romaenko YuV, Speransky KE, Fedynsky AV (1981) Atmos-opt yablen nabluch orbit nauch stanchii 'salyut', Tartu pp 46–51 (in Russian)

Witt G (1960) Polarization of light from noctilucent clouds. J Geophys Res 65:925-933

Witt G (1962) Height, structure and displacements of noctilucent clouds. Tellus 14:1-18

Witt G (1968) Optical characteristics of mesospheric aerosols distributions in relation to noctilucent clouds. Tellus 20:98-114

Witt G (1969) The nature of noctilucent clouds. Space Res IX:157-169

Witt G, Wilhelm N, Stegman J, Williams AP, Holback B, Llewellyn E, Pedersen A (1971) Rep AP-7 Inst Meteorol Univ Stockholm

Witt G, Dye JE, Wilhelm N (1976) Rocket-borne measurements of scattered sunlight in the mesosphere. J Atmos Terr Phys 38:223-238

WMO (1970) International noctilucent cloud manual, WMO No. 250.TP.138, Geneva

Zakharov VN (1988) Photogram measurements of apparent motions of noctilucent clouds. Issl po geomagn aeron i fiz col 80:69-78 (in Russian)

Zhukova LP, Trubnikov BM (1966) On the dynamics of meso-scale wave agreements in the field of noctilucent clouds and the penetration of the troposphere meso-scale waves into the upper layers of the atmosphere. Proc noctilucent clouds, Tallinn pp 222-231

Note added in proof:

Jensen EJ, Thomas GE, Toon OB (1989) On the diurnal variation of noctilucent clouds. J Geophys Res 94: in press

Garcia RR (1989) Dynamics, radiation and photochemistry in the mesosphere implications for the formation of noctilucent clouds. J Geophys Res 94 (in press)

Appendix 1: Atmospheric Refraction

The problem of allowing for refraction through a medium that has a stratified variation in refractive index and spherical symmetry has been studied for centuries. Kepler (1604) took the empirical data of Tycho Brahe and constructed separate tables of atmospheric refraction for the Moon, the Sun and stars. (Kepler did not have Snell's law for the refraction of light passing from one optical medium to another available to him.).

The solution of atmospheric refraction is usually given as a table of values of apparent altitude and true zenith distance for a star, together with air mass corresponding to the specified ray (for example, see Allen 1963). More elaborate tables, allowing the calculation of refraction for a model atmosphere which gives tractability in the mathematical analysis, are available (e.g. Garfinkel 1944). None of these is very convenient for twilight calculations and, in general, it is better to do a numerical integration using some particular model atmosphere.

The integration has some difficulties, however. The differential equation is obtained by applying Snell's law to a spherically-symmetric, horizontally stratified atmosphere (see Fig. A1.1) which leads to

$$N (R_E + H) \sin I = \text{a constant, } C. \tag{A1.1}$$

Here, N is the atmospheric refractive index at height H above a spherical Earth of radius R_E. I is the angle of incidence of the stellar ray where it passes the layer at height H.

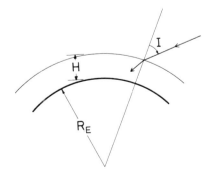

Fig. A1.1. The geometry of atmospheric refraction

The path of the ray is the locus of the points (r, ϕ) which are solutions of the equation

$$r\, d\phi/dr = C/\sqrt{N^2R^2 - C^2}. \tag{A1.2}$$

In the twilight problem, the path of a "grazing ray" is needed (see Fig. A1.2): such a ray is tangent to the stratified layers in the observer's zenith, that is, it passes horizontally above the observer's head. At the tangent point, $I = 90°$ and $C = Nr$. Consequently, $(d\phi/dr)$ becomes infinite and the numerical integration fails.

After a number of trials of different methods of numerical integration for a ray entering the atmosphere, the Runge-Kutta fourth-order integration was found to give an acceptably convergent solution as the step size $(\Delta\phi)$ was decreased. Using double precision in the computer (2×36-bit word size), rounding errors became appreciable when $\Delta\phi$ was less than 0.00003. The balance between discretization error and rounding error seems to be held for $\Delta\phi = 0.00005$ radians. There remain noticeable quantization errors in the results. Integration along an incoming ray was finally discarded in favour of outward integration. The initial conditions were a chosen minimum height, Z_M, and $I = 90°$; the first integration step was avoided by calculation of the right-angled triangle using the step size, $\Delta\phi$, chosen for subsequent integration. Euler's rule was chosen for this integration, carrying the calculation for three step sizes, respectively one, two and three times $\Delta\phi = 0.00008$ with quadratic extrapolation of the results back to zero step size.

It is clear (Fig. A1.2) that the so-called angle of refraction for the computed ray is simply the quantity β and that this is related to the height of the intercept of the projected (unrefracted) ray in the observer's zenith through the equation

$$\cos\beta = C/(R_E + Z) \tag{A1.3}$$

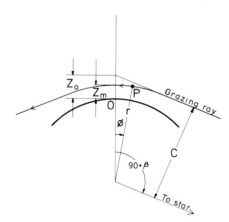

Fig. A1.2. The geometry of the "grazing ray" passing at height Z_M above an observer at O. The ray comes from a star lying at an angle β below the observer's horizon

and, from above,

$$C = N_M(R_E + Z_M),$$ \hfill (A1.4)

where N_M is the refractive index at height Z_M.

The model atmosphere used is listed in Table A1.1, which gives the molecular number density and temperature for 1 km height intervals. Table A1.2 lists the calculated β for set values of C (with a ground level refractive index of 1.000282, C = 6371.8 km for Z_M = 0 km). There is an appreciable optical dispersion of the grazing rays, and values of β, Z_M are listed in Table A1.3 for the first 15 rays listed in Table A1.2.

Table A1.1. Molecular number density and temperature in the model atmosphere used in tracing rays. Each set of numbers is for altitudes from 0 to 100 km at 1-km intervals

Number density (m^{-3}):

2.545 + 25	2.312 + 25	2.093 + 25	1.891 + 25	1.704 + 25
1.531 + 25	1.373 + 25	1.227 + 25	1.093 + 25	9.712 + 24
8.593 + 24	7.585 + 24	6.486 + 24	5.543 + 24	4.738 + 24
4.050 + 24	3.461 + 24	2.956 + 24	2.529 + 24	2.162 + 24
1.849 + 24	1.574 + 24	1.341 + 24	1.144 + 24	9.760 + 23
8.335 + 23	7.123 + 23	6.092 + 23	5.214 + 23	4.466 + 23
3.778 + 23	3.237 + 23	2.774 + 23	2.383 + 23	2.049 + 23
1.764 + 23	1.521 + 23	1.313 + 23	1.135 + 23	9.821 + 22
8.514 + 22	7.385 + 22	6.413 + 22	5.576 + 22	4.855 + 22
4.233 + 22	3.694 + 22	3.227 + 22	2.820 + 22	2.469 + 22
2.164 + 22	1.931 + 22	1.720 + 22	1.531 + 22	1.361 + 22
1.208 + 22	1.071 + 22	9.484 + 21	8.388 + 21	7.409 + 21
6.535 + 21	5.754 + 21	5.059 + 21	4.442 + 21	3.893 + 21
3.408 + 21	2.980 + 21	2.601 + 21	2.267 + 21	1.971 + 21
1.712 + 21	1.487 + 21	1.289 + 21	1.115 + 21	9.622 + 20
8.291 + 20	7.127 + 20	6.108 + 20	5.227 + 20	4.459 + 20
3.795 + 20	3.173 + 20	2.654 + 20	2.219 + 20	1.856 + 20
1.553 + 20	1.299 + 20	1.087 + 20	9.097 + 19	7.615 + 19
6.369 + 19	5.267 + 19	4.520 + 19	3.621 + 19	3.018 + 19
2.523 + 19	2.116 + 19	1.742 + 19	1.500 + 19	1.271 + 19
1.077 + 19				

Table A1.1 *(Cont.)*

Temperature (K):

293	293	286	280	273	266	259	252	246	240
233	224	217	215	209	203	204	205	206	208
210	213	216	218	222	225	228	231	234	237
240	243	246	249	251	254	257	260	262	265
268	272	275	277	280	283	284	285	287	287
287	288	288	286	285	284	281	277	274	269
264	259	253	249	244	238	233	229	223	219
214	209	203	199	194	189	184	180	176	173
169	166	163	161	159	157	156	156	156	156
157	160	165	169	174	178	183	187	192	197
200									

Table A1.2. Atmospheric refraction for grazing rays; wavelength $(\lambda) = 590$ nm

Constant (km)	Angle of refraction β (radians)	Height of grazing ray Z_M (km)	Height of intercept Z_0 (km)
6372	0.009192	0.30	2.27
6373	8252	1.48	3.22
6374	7398	2.65	4.17
6375	6636	3.80	5.14
6376	5945	4.93	6.11
6377	5340	6.06	7.09
6378	4776	7.17	8.07
6379	4261	8.26	9.06
6380	0.003815	9.35	10.05
6381	3433	10.43	11.04
6382	3126	11.51	12.03
6383	2637	12.59	13.02
6384	2222	13.65	14.02
6385	1884	14.70	15.01
6386	1600	15.75	16.01
6387	1354	16.79	17.01
6388	1149	17.82	18.00
6389	978	18.85	19.00
6390	0.000836	19.87	20.00
6391	710	20.89	21.00
6392	600	21.91	22.00
6393	509	22.92	23.00
6394	432	23.93	24.00
6395	367	24.94	25.00
6396	312	25.95	26.00
6397	266	26.96	27.00
6398	227	27.96	28.00
6399	197	28.97	29.00
6400	0.000164	29.97	30.00

Table A1.3. Dispersion of the lowest grazing rays

C (km)	$\lambda = 380$ nm β (radian)	Z (km)	$\lambda = 800$ nm β (radian)	Z (km)
6372	0.009473	0.25	0.009113	0.31
6373	8498	1.44	8182	1.49
6374	7613	2.61	7337	2.66
6375	6825	3.77	6583	3.81
6376	6110	4.90	5898	4.94
6377	5488	6.03	5298	6.06
6378	4906	7.14	4738	7.17
6379	4375	8.25	4227	8.27
6380	0.003917	9.34	3786	9.36
6381	3522	10.42	3408	10.44
6382	3209	11.50	3102	11.51
6383	2706	12.58	2617	12.59
6384	2279	13.64	2206	13.65
6385	1932	14.70	1870	14.70
6386	1640	15.74	1588	15.75

Appendix 2: Atmospheric Transmission Along Grazing Rays

The path of a ray can be calculated quite precisely using the method outlined in Appendix 1. Clearly, the optical transmission along the path can then be estimated given a model atmosphere. Due to the variation of the angle of refraction, β, with change in grazing height, Z_M (see Fig. A1.2), rays that enter the atmosphere parallel will leave the atmosphere diverging from one another. The dilution of the transmitted radiation must be allowed for. To illustrate the magnitudes involved, the transmission of a number of rays has been calculated for the model atmosphere outlined in Appendix 1. The molecular number density given there can be used to estimate the amount of light scattered from the ray by molecular scattering. In addition, as has been pointed out in Chapter 3, adsorption in the stratosphere by ozone is of great importance under twilight conditions. For demonstration purposes, the model ozone layer listed in Table A2.1 has been used, together with the Chappuis absorption coefficients measured by Vassy (1941) and listed in Table A2.2.

Table A2.3 gives the atmospheric transmittance calculated for some grazing rays passing right through the atmosphere. Dilution in the amount appropriate for sunlight falling on a noctilucent cloud layer at 82-km height has been included, but there is no account taken of the solar disc, or its limb darkening. The model atmosphere does not contain aerosols, clouds or particulate matter that may be present in a real case and for which allowance would then be needed.

Table A2.1. Ozone layer used in the atmospheric model. The ozone number density (m^{-3}) is given for the height interval 0 to 59 km at 1-km intervals. The total vertical amount of ozone is 0.312 cm at STP

Ozone number density $\times\ 10^{-18}\ m^{-3}$

0.24	0.24	0.24	0.24	0.24	0.24	0.24	0.46	0.96	1.40
1.84	2.28	2.73	3.20	3.72	4.18	4.66	5.02	5.05	5.02
4.88	4.67	4.36	3.95	3.60	3.28	2.90	2.52	2.18	1.82
1.41	1.08	0.89	0.73	0.60	0.49	0.39	0.32	0.27	0.22
0.18	0.15	0.13	0.11	0.09	0.08	0.07	0.06	0.04	0.03
0.03	0.02	0.01	0.01	0.01	0.00	0.00	0.00	0.00	0.00

Table A2.2. Chappuis band optical density (after Vassy, 1941)

λ (nm)	k_λ (cm^{-1})	λ (nm)	k_λ (cm^{-1})	λ (nm)	k_λ (cm^{-1})
440	0.0005	550	0.0450	660	0.0281
450	0.0014	560	0.0521	670	0.0220
460	0.0039	570	0.0607	680	0.0177
470	0.0039	580	0.0607	690	0.0143
480	0.0098	590	0.0584	700	0.0107
490	0.0098	600	0.0678	710	0.0078
500	0.0166	610	0.0635	720	0.0064
510	0.0201	620	0.0515	730	0.0041
520	0.0246	630	0.0447	740	0.0031
630	0.0348	640	0.0378	750	0.0035
540	0.0384	650	0.0313	760	0.0019

Table A2.3. Atmospheric transmittances for the 15 lowest grazing rays; the constant C is incremented in 1–km steps, from 6372 km to 6386 km

λ (nm)	6372 km	6373 km	6374 km	6375 km	6376 km
400	0.000	0.000	0.000	0.000	0.000
450	0.000	0.000	0.000	0.000	0.000
500	0.000	0.000	0.000	0.000	0.001
550	0.000	0.000	0.001	0.002	0.003
600	0.001	0.001	0.002	0.004	0.006
650	0.006	0.010	0.016	0.024	0.034
700	0.026	0.038	0.054	0.073	0.099
750	0.058	0.080	0.106	0.136	0.173
800	0.102	0.131	0.166	0.203	0.248

λ (nm)	6377 km	6378 km	6379 km	6380 km	6381 km
400	0.000	0.000	0.000	0.000	0.000
450	0.000	0.000	0.001	0.003	0.008
500	0.002	0.004	0.008	0.014	0.024
550	0.005	0.008	0.013	0.018	0.025
600	0.009	0.012	0.015	0.018	0.023
650	0.046	0.058	0.071	0.087	0.105
700	0.125	0.152	0.184	0.219	0.260
750	0.209	0.247	0.291	0.337	0.391
800	0.289	0.333	0.384	0.437	0.497

λ (nm)	6382 km	6383 km	6384 km	6385 km	6386 km
400	0.001	0.004	0.009	0.019	0.034
450	0.013	0.026	0.046	0.074	0.108
500	0.032	0.050	0.073	0.100	0.132
550	0.029	0.039	0.051	0.064	0.079
600	0.023	0.028	0.034	0.041	0.049
650	0.105	0.124	0.146	0.167	0.189
700	0.255	0.295	0.338	0.377	0.415
750	0.375	0.426	0.479	0.526	0.568
800	0.470	0.525	0.583	0.632	0.674

Subject Index

Abnormal Observations   13, 90, 106–107
Aerosol chemistry   28, 108f, 114–115
Artificial noctilucent clouds   105–106
Atmospheric emission   99–100
    airglow   13–14, 46–47, 51, 56, 90, 98,
       104, 126, 128
    atmospheric models   10, 11, 94, 122, 151
Aurora borealis and noctilucent clouds   1,
    10, 13, 45–46, 90, 98–99, 100, 101, 103
Aurora pseudo   13

Belts, or zones of noctilucent clouds   28, 56,
    98
Berlin Atmospheric Programme   16–22, 36
Breakdown of summerly mesosphere   95

CAMP   22, 111

Daily variation of noctilucent clouds   87, 96
Displays, on nights
    1885, June 8   15
    1885, July 19   37
    1885, July 22   37
    1886, June 22   31
    1886, July 27   44
    1889, July 2   17, 31, 37
    1889, July 9   37
    1889, July 24   31
    1889, July 31   31
    1890, July 6   31, 37
    1890, July 9   37
    1890, July 10   31, 37
    1891, June 25   31
    1894, July 12   37
    1908, June 30   39
    1932, July 10   31
    1932, July 24   31
    1934, June 30   31, 39
    1949, July 10   31
    1950, July 24   98
    1951, July 18   37
    1953, July 12   37
    1958, August 10   30–31, 40–41, 65
    1958, June 16   31
    1959, July 30   66

    1959, July 14   31
    1960, July 21   37
    1960, July 20   31
    1960, August 5   31
    1961, June 30   31
    1962, June 7   37
    1963, July 27   37
    1963, July 30   93, 100
    1964, June 6   37
    1964, June 30   37
    1964, July 16   37
    1965, July 7   37
    1965, July 15   31
    1965, July 19   31
    1965, July 24   37
    1967, July 3   39
    1967, July 16   31
    1967, August 9   31
    1969, July 22   45
    1970, July 10   47
    1972, June 14   46
    1974, June 27   7
    1974, July 9   46
    1974, July 18   6
    1974, July 23   29
    1975, June 29   67
    1976, June 18   38
    1979, July 10   31
    1981, March 22   106
    1981, May 14   106
    1981, July 31   106
    1984, June 28   63
    1984, June 29   64
    1986, July 23   1, 5, 8–9
D-Region   88, 95, 110–111
Drift motions of noctilucent clouds   36ff

Fe   110
FeO   110
FeSi   110, 120

Geographical distribution
    of noctilucent clouds   1–3, 4, 27–28
    northern hemisphere   53–56, 89–90, 128
    southern hemisphere   56, 91, 107

Geometry of observation   23–24, 30, 32–34
Glow   13–14
Gravity waves   4, 40–42, 121, 128, 130
Grenade soundings   92

Height of noctilucent clouds   16, 18, 26, 30–
   32, 34–36, 42–43

Kp-Indices
   relations with noctilucent clouds
      88, 100–103, 111
Krakatoa eruption   1, 14–16, 22, 125

Laser observations   105, 114
Lidar observations   104f
Lunar effects, possible connection   104f

Magnetic storms, possible relationship   88,
   100ff
Maxwell equations   62
Measurements of noctilucent cloud pates
   30ff, 35f
   formulae   23–24ff, 35f
   graphical methods   24ff
   identification of stars   34ff, 30ff, 3ff
   single baseline   26, 30, 35, 39, 71, 101, 106
Mesopause   11, 93–94, 93, 98, 109, 113, 120–
   123
Mesosphere   11, 92–94, 100, 103, 112–113,
   121
   circulation system   12, 94–96, 98, 110,
      112
   mixing ratio   2, 113, 117
   models   92, 94
   temperature   3, 11, 94–96, 100, 108, 111,
      114–116
   transition periods   94–96
   water vapour   112, 117
   wind   84, 92, 95, 110, 112
Meteoric currents   7, 13–14, 125
Meteoric dust   91–92, 112, 114, 120
Meteoric ions   7, 111–112, 120
Meteoric smoke particles   122–124
MgO   120
MgO-FeO   120
Mie scattering theory   53–54, 65–66
Mother of pearl clouds   18, 40
Movement of noctilucent clouds   18–36, 43
Moving pictures of noctilucent clouds   35–36

NO⁺   110–111

Observational methods of noctilucent clouds
   1–12, 15–22, 23ff, 34–36, 75ff, 87, 89ff
OGO-6   49, 51

Optical methods of measuring height and
   position   25ff, 30ff, 94f
Ozone   2, 10–11, 26, 29–30, 43, 52, 70, 113,
   116
   chappius band   29, 45, 70
   ozone absorption   42, 52, 70
   ozone layer   2, 26, 43, 70

Parallactic photography   35, 106
Particles   33–36, 44, 46ff, 48, 74, 77–78, 80–
   81, 115–117, 120
   cosmic dust   58, 84, 91–92, 120
   growth of particles   54ff, 74, 115–116,
      120ff, 124, 127
   meteoric   44
   nucleation of particles   47, 112, 114, 118f,
      122f, 124
   size distribution   44, 74, 78–79, 84, 86, 110,
      123
Photographic work in
   Canada   20, 36, 89–90, 100
   Chile   91
   Germany   4, 22, 36, 89–90, 101
   Poland   4
   Scandinavia   19, 30, 36, 40, 65, 89
   United Kingdom   4, 87–90, 99–102
   USA   20, 27, 36, 40, 89, 101–102
   USSR   4, 18, 20, 36, 89–90
Photographs of noctilucent clouds in colour
   6–9
Plates, suitable for noctilucent clouds   4–5,
   33–36
Polar mesospheric clouds   22, 54, 128–129
   relationship with noctilucent clouds   54,
      127–130
Polar vortex   94
Polarization   30, 44, 58–74
   circular   67–74
   degree of   58
   state of   58

Radar   121
Radiation   114ff
Rayleigh scattering   30, 52, 60–61, 67–68
Refraction   65, 149ff
Reliability of noctilucent cloud data   89
Rocket   11, 31, 35–36, 46–48, 78–82, 85, 93–
   96, 109–111
   flights by the Max-Planck-Institute Heidel-
      berg 1968/70   85 ff
   flights over Canada 1968/70   82–85
   flights over Sweden 1962/67   76–80
   flights over Sweden 1970/71   81–82
   rocket measurements   75ff, 95–96
   rocketborne photometer   31, 46, 109–111
   rocketborne sampling   31, 57, 75, 111

Satellite data of noctilucent clouds   22, 50–57, 106, 115, 127–128
Seasonal variation of noctilucent clouds
  northern hemisphere   3–4, 27, 87, 89–91, 97
  southern hemisphere   3, 91, 107
Solar activity and noctilucent clouds   12, 22, 87–88
Solar depression angle   23–25, 27
Solar heating   98–102
Solar Mesosphere Explorer (SME)   22, 51–55, 127–128
Spectrophotomtery by rockets   29, 46–47, 50, 75
  by satellites   50–54, 56
Spectrum of noctilucent clouds
  measurements   29, 43ff, 54ff, 99, 110
  observations   44, 54ff, 99, 110
  typical   43 ff
Stokes parameter   58–62, 70, 73
Stokes vector   58–62

Temperature
  mesopause   11, 22, 92–96
  mesosphere   11, 22, 93–96
  rocket data   11, 22, 92–96
Theories of noctilucent clouds   18–19, 56, 74, 78, 86–87, 122, 124ff, 127–129
Thickness of noctilucent clouds   73, 104–105
Transition periods, relation with noctilucent clouds   12, 91, 94–96
Tropical noctilucent clouds   106
Troposphere jet stream   41
Types of noctilucent clouds   4–10, 34, 38–42

Velocity of noctilucent clouds   18, 36–38
VHF incoherent scatter spectra   111, 121
Volcanic, origin of noctilucent clouds   15–19, 124-125
Volcanic dust   15–19, 124–125
Volcanic relationship   15–19, 125–126

Wave structure of noctilucent clouds   12, 38ff, 41–42, 117
Winds
  mesopause   36–41, 93, 96
  mesosphere   12, 36, 39, 92–96
Winter data of noctilucent clouds   90

Name Index

Alekseev AM 134
Allen CW 134, 149
Anderson JG 99, 113, 134
Andre L 139
Andrews JG 134
Apruzese JP 134
Arago F 13, 131
Arakawa H 134
Archenhold FS 15–17, 19, 36–37, 80, 89, 132
Archenhold G 91
Arnold F 110, 113, 131, 134–135, 141
Asano S 62, 134
Astapowitsch IS 88–89, 123, 132, 134
Austin J 1, 134
Avakyan S 106, 134
Avaste, OA 26, 55–56, 134

Backhouse TW 15, 100, 131
Baggaley WJ 14, 134
Baibulatov 115, 134
Bailey AD 143
Ball S 95, 147
Balsley BP 121, 134, 143
Bandeen WR 146
Bandermann 62, 134
Banks PM 98, 134–135
Barat J 135
Barber P 42, 62, 135, 142
Barkow E 132
Bartman FL 146
Bary E 47, 135
Batten ES 42, 135
Battermann H 132
Bauernberger H 132
Beeson DE 31, 47–48, 57, 73, 147
Belton MJS 135
Belyaev BI 55, 57, 135
Benech B 105, 135
Bennett JM 138
Bernhardt KH 22, 138
Bertin F 141
Bessonova TD 88–89, 135
Bevilacqua RM 113, 115, 135, 138, 144
Bezrukova AY 88, 135
Bischof M 140

Björn LG 110–111, 141
Blamont JE 49, 51, 135, 137
Blanchard MB 137
Blum PW 94–95, 138, 143
Böhme W 135
Bohren F 74
Bologna JM 135
Borbely E 96, 135
Brigg EK 135
Brodhun D 41, 144
Bronshten VA 13, 19, 36–37, 114, 132, 135
Brown TJ 135
Burov MI 31, 36–37, 135–136
Busch F 132
Byrne FD 100, 136

Carter DM 143
Chanin ML 105, 136
Chao J 136
Chapman S 10, 42, 132, 136
Charlson RJ 42, 125, 136
Charney JG 42, 136
Chrest SA 146
Christie AD 20, 89, 92, 136
Church MJ 146
Chvostikov IA 19, 22, 42, 125, 136, 143
Clairemidi J 98, 136
Clarke D 60, 136
Croskey CL 144
Crutzen JP 146
Currie BW 20, 136

D'Angelo N 88, 101–102
Danilov AD 42, 136
Deguchi S 113, 136
Deirmendijan D 45, 57, 114, 125, 136, 147
Della Luca L 143
Deluce F 13, 131
Dessens J 105, 135
Dickinson PHG 141
Dickinson RE 136
Dieminger W 136
Dirikis MA 31, 36–37, 136–137
Donahue MT 22, 49–50, 57, 99, 112–113, 134, 137, 142

Drazin PG   42, 136
Dubin M   137
Dye JE   72, 148

Ebel A   92, 94, 96, 137, 139–140
Eberhardt P   141
Eckart C   41, 137
Ecklund WL   134
Edwards HD   36, 145
Eerme KA   134, 147
Einstein A   121, 132
Eisenberg D   116, 137
Entzian G   95, 142, 147
Epstein PS   121, 132
Ertel H   13, 41–42, 87, 126, 132, 137
Evdokimento SV   137

Farlow NH   82–84, 137, 147
Fast NP   19, 106, 137
Fast WH   106, 137
Fechtig H   85, 137, 144
Fedynski AV   133, 137, 147
Fehsenfeld FC   146
Ferguson EE   137, 146
Ferry GV   82–84, 137
Fessenkov GV   133, 137
Feuerstein M   85, 137
Fiare AC   144
Fiocco G   48, 87, 105, 114–115, 137–138
Förster W   13, 16, 18, 22, 80, 125, 131, 133
Fogle B   3, 27, 36–37, 40–41, 45–46, 48, 51,
   57, 87–89, 91, 101, 126, 138
Francis RJ   138
Franzmann YL   137
Frings WG   139
Fritt DC   134
Fryklund DH   144
Fullam EF   140

Gärtner V   94, 138, 143
Ganguly S   13, 111
Garcia RC   42, 137, 138
Garfinkel B   133, 149
Gavine DM   1, 4, 89, 138
Geller MA   138
Gerloff U   137
Germashevsky M   134
Giauque WF   116
Gibbins JJ   122, 138, 144
Glöde P   138
Götzelmann A   138
Goldberg RA   110, 139
Gosshard EE   137, 139
Gottlieb CB   142, 144
Gottlieb EW   144
Grahn S   31–32, 139
Grainger JF   60, 136

Grams G   48, 87, 105, 114, 137–138
Grechko GM   26, 50, 134
Greenhow JS   139
Gregory JB   139
Griffiths L   107, 139
Grishin NI   13, 39, 42, 45, 57, 135, 139
Grjebine T   139
Gromova TD   88–89, 139
Grossmann KU   111, 139
Groves GV   93–94, 139
Gruner P   133
Guenther B   49, 57, 137
Guthnick M   139

Hale LC   144
Hallert B   35, 139
Hallgren DS   81–82, 139–140
Hallissey M   4, 89, 143
Hamilton RA   90, 139
Hanson AM   139
Hapgod MA   34, 139, 146
Harrison AW   46, 57, 100, 139
Hartmann GK   113, 139
Hartmann W   133
Hartwig E   131
Hass H   139
Haurwitz B   40–41, 88, 126, 133, 139, 144
Heard WC   139
Heintzenberg   73–74, 139
Helmholtz Hv   41, 131
Helmholtz Rv   131
Hemenway CL   68, 76, 78–79, 81, 139–140,
   142
Hermann U   135
Hersé M   13, 130, 136, 143
Hesstvedt E   92, 125, 140
Heymsfield AJ   140
Hines CO   40–41, 140
Hobbs PV   118, 140
Hoffmeister C   13, 37, 90, 133, 140
Holback B   148
Hollenbach D   147
Holton JR   92, 134, 140
Houghton JT   2, 92, 140
Howard RJ   135
Hulburt EO   138
Hulst VD   140
Humi M   146
Hummel J   50–52, 57, 140
Humphreys WJ   20, 133
Hunt BG   40, 110, 140
Hunten DM   112, 140

Ivaniya SP   115, 134
Ivanovsky AI   140
Iwasaka Y   125, 140
Izakov MN   115, 140

Jager C   135
Jakobs JH   42, 137, 140
Jardetzky W   19, 125, 133
Jensen EE   22, 96, 127–129, 140, 141, 148
Jesse O   13, 15–19, 25, 26, 30–31, 36–38, 44,
    89, 91, 131
Johannessen A   110, 140
Johnson FS   141
Joiner RG   144
Joos W   110, 134
Joseph JH   141
Juskeseleiva L   141

Kaiser TR   141
Kallmann-Bijl KH   141
Kashtanov AF   122, 141
Katchen L   143
Katchen W   146
Kato S   141
Kauzmann B   116
Kauzmann H   137
Keegan TJ   42, 141
Keese RG   147
Keevallik   55
Kellogg WW   141
Kemp JC   62, 135
Kendall PC   42, 136
Keng EYA   122, 144
Kepler J   131, 149
Kerker M   60, 61, 122, 141, 146
Kerridge JF   140
Khantodze AG   141
Kießling J   14, 125, 131–132
Kinmark I   139, 141
Kiselevsky LI   135
Kleinert H   133
Klimuk PI   134, 135, 147
Kochanski A   92, 141
Kockarts G   135
Kohl G   42, 141
Kokin GA   140
Koksharov PI   135, 147
Konasenok VN   141
Kopp E   111, 139–140
Kosibowa W   4, 141
Kovalyono KV   106, 134
Krankowsky D   110, 113, 134, 139–141
Krassovsky VI   141
Kropotkina EP   104, 141
Kuhnke F   41, 142
Kunzi, FK   139
Kurilova JV   39, 139, 142

Latimer P   62, 142
Lauter EA   88, 94–95, 142
Lazarev AI   49, 134, 142

LeConte J   132
Leonov AA   49, 142
Leovy C   42, 92, 134, 142
Leslie RC   15, 132
Lilley EA   142, 144
Lindzen RS   142
Linscott J   79, 142
Litvak MM   142, 144
Liu SC   112, 113, 142
Llewellyn E   148
Lloyd KH   142
Lobsiger E   139
Lovchikova LP   135
Low CH   142
Lowan AN   142
Ludlam FH   26, 36, 142
Lundin A   135

Mackinnon DJ   135
Maignon E   13, 132
Mairan JJ   13, 132
Malzev V   22, 133
Mannig L   146
Marron M   13, 132
Martin H   142
Martynkevic GM 42, 142
McDonald JE   142
McGovern WE   146
McIntosh DH   4, 34, 89, 142–143, 146
McKay PC   22, 51–54, 57, 122, 128, 143, 147
McLone RR   143
Megrelishvili TG   42, 136, 143
Meier L   37, 143
Meinel AB   105, 143
Memmesheimer M   92, 94, 95, 138, 143
Middlehurst B   143
Mie G   35, 133
Millmann PM   20, 143
Minnaert M   3, 143
Mohn H   20
Mohnen V   139
Mookins EE   36, 136
Moreels G   136, 143
Morozov VM   143
Muhleman DO   113, 136
Munk W   139
Murgatroyd RJ   143

Narcisi RS   143
Nastrom GD   143
Nestorov H   143
Neufeld EL   139
Nicolet M   113, 143
Nordberg WG   38, 143, 146
Novikov BM   122, 141
Novozilov NJ   143

Offermann D   139
Oleak H   144
Olivero JJ   22, 51, 52, 54, 113, 115, 135, 140, 144, 147
Omond RT   132
Ordt Auffm N   41, 144
Orr C   122, 144

Packer DM   49, 144
Packer IG   49, 144
Pas'ko LN   134
Paton J   4, 20–22, 24–27, 31, 79, 87, 89–90, 98–99, 100–102, 133, 144
Paulson JF   147
Pavlova TD   144
Pedersen A   148
Penning E   132
Pepper W   133
Pernter JM   15, 125, 132
Perov SP   140
Peters CFW   13, 132
Philbrick CR   111, 141, 144
Pierraard JM   145
Plyuta VE   135
Pokrovskii KD   132
Pollermann B   144
Popov OI   147
Pröll G   132
Pyka JL   4, 140, 141

Quiroz RS   36, 38, 144

Radford HE   113, 144
Rajchl J   144
Rampino MR   146
Rauser F   85, 137, 144
Rayleigh Lord   61, 132
Reddy CA   41, 140
Rees MH   45–46, 48, 51, 57, 138
Reid GC   110, 120–121, 144
Reiter ER   41, 144
Rey TR   146
Ricketts WB   135
Roddy AF   45, 120, 145
Rössler F   31, 47, 57, 135, 145
Romanenko YUV   134, 147
Roper RH   145
Rosenberg NW   36, 145
Rosenfeld SHK   40, 121, 127, 145
Rosenthal SK   142
Rosinski J   145
Rothwell P   138
Rowan DJ   132
Rozenberg GV   29, 145
Rudzki MP   133
Ryon TG   146

Sacher P   146
Sargin VJA   146
Savinykh V   106, 134
Scherhag R   94, 145
Schilling GF   141
Schmalberger DC   139
Schmidt A   133
Schmidt J   13, 132
Schmitz G   145
Schoeberl MR   134
Schulte P   146
Schwartz PR   135, 138, 144
Scott AFD   42, 146
Scultetus H   133
Seedsman DL   138
Seleznev GI   147
Self S   146
Semenova VI   147
Sergeyevich VN   135, 147
Sergin SJA   146
Seshamani R   102, 146
Sevastyanov VI   134–135, 147
Shafrir U   146
Shanklin L   139
Sharonov VV   87, 146
Shaw C   132
Shefov NN   99–100, 104, 129, 141, 146
Sherman C   143
Shinjo S   133
Shurcliff WA   58, 146
Simmons DAR   25, 34, 38, 89 146
Singleton F   143
Sitarski M   122, 146
Skrivanek RA   140, 146, 160
Skuratova IS   42, 147
Smith D   146
Smith WS   93, 95–96, 109, 143, 146
Smullin LD   137
Smyth CP   1, 44–45, 100, 132
Soberman RK   68, 139, 140, 146
Solomon S   42, 112, 138, 144, 146
Speil J   141
Speransky KE   147
Speth P   137, 140
Sprenger K   142
Szrenewskii B   132
Stegman J   148
Stenzel A   133
Störmer C   13, 20–21, 30–31, 35, 39, 133
Strobel, DF   134
Stroud, WC   143, 146
Stut JW   116, 143, 146
Süring R   41, 133
Sved GM   141
Swartz PC   146
Symmons GJ   14, 125, 132

Tackett DC   139
Tarasova TM   66, 146
Taubenheim J   42, 95, 146–147
Taylor MJ   25–26, 31, 34–35, 42, 71, 138–139,
   146
Thacker DL   135, 138, 144
Theon JS   38, 91, 92, 94–96, 110, 143, 146
Thomas DM   51, 143
Thomas GE   22, 51–54, 57, 96, 127–130,
   140–144, 147, 148
Thomas L   110, 147
Thomas TF   113, 147
Thrane EV   110, 141
Titus P   146
Toon OB   140, 147, 148
Tozer R   31, 47–48, 57, 73, 147
Trubnikov BN   42, 147–148
Tseraskij VK   15, 132
Tsou IJ   144
Turco RP   124–125, 127, 129, 140, 149
Twelves S   147

Ungestrup E   88, 101–102

Vasil'ev NV   19, 147
Vasilyev OB   66, 88, 147
Vassy A   133, 154
Vegard L   14, 133
Veismann UK   134
Verbeck RDM   14, 132

Verdeil F   13, 132
Veselov DP   46, 147
Vestine EH   20, 36, 45, 57, 87, 114, 125, 133,
   136, 147
Vincent RA   95, 138, 142, 147
Visconti G   125, 138, 147
Von Cossard GV   147

Wait JR   61, 147
Webb W   125, 147
Wegener A   19, 41, 133
Wegener K   133
Whitten RC   147
Wilhelm N   72, 146, 148
Williams AP   148
Willmann CH   49, 66–69, 88, 115, 134, 147
Wilson WJ   135
Witt G   30–32, 36–37, 40, 46–47, 57, 58, 65–
   66, 72–73, 80, 94, 108, 135, 139–140, 142,
   147–148
Wörner H   133
Wolf M   14, 133

Yamamoto   62, 134
Yanovsky AF   135
Yeh C   62, 135

Zakharov VN   148
Zatejssikov G   36, 37
Zhukova LP   40, 147